# REGULATION OF DIFFERENTIATION IN MAMMALIAN NERVE CELLS

# REGULATION OF DIFFERENTIATION IN MAMMALIAN NERVE CELLS

## Kedar N. Prasad

*University of Colorado Medical Center*
*Denver, Colorado*

Plenum Press   ·   New York and London

Library of Congress Cataloging in Publication Data

Prasad, Kedar N
  Regulation of differentiation in mammalian nerve cells.

  Includes index.
  1. Cell differentiation. 2. Neurons. 3. Cellular control mechanisms. 4. Mammals—
Cytology. I. Title. [DNLM: 1. Neurons—Cytology. 2. Cell differentiation. 3. Mammals—
Embryology. WL 102.5 P911r]

| | | |
|---|---|---|
| QH607.P73 | 591.8′761 | 79-18944 |

ISBN 978-1-4684-8114-3     ISBN 978-1-4684-8112-9 (eBook)
DOI 10.1007/978-1-4684-8112-9

© 1980 Plenum Press, New York
Softcover reprint of the hardcover 1st edition 1980

A Division of Plenum Publishing Corporation
227 West 17th Street, New York, N.Y. 10011

Dedicated to my sister, Kamla Devi

# PREFACE

Several model systems have been used to understand the cellular and molecular mechanisms of differentiation of mammalian nerve cells. Each model system has unique advantages and disadvantages and is suited for the study of only certain aspects of differentiation. In this book, the techniques of these models and the usefulness and limitation of each model system are discussed.

An awareness of the use and misuse of each model system is important for a rational interpretation of data and for a reasonable comparison of data obtained from different model systems. With the use of clonal lines of neuronal cells and hybrid neural cells (neural cells × nonneural cells), many new concepts have emerged concerning the regulation of differentiated functions, the relationship between the expressions of individual differentiated functions, and the relationship between differentiation and malignancy. Some of these concepts have already proved to be relevant to regulation of differentiation *in vivo*. These new emerging concepts are discussed extensively in this book.

Many new agents (physiological and nonphysiological) which induce or increase the expression of one or more differentiated functions have been identified. These agents will be useful biological tools for further studies of the regulation of differentiation in mammalian nerve cells. This book describes the role of each agent in differentiation of nerve cells by focusing on different model systems and provides a rational basis for selecting the particular differentiating agents for specific problems of differentiation processes.

Also contained in this volume is recent information on the mechanism of neural induction and the role of cAMP in such processes. Furthermore, the role of cAMP, nerve growth factor, and other agents in regulating the expression of differentiated functions is discussed extensively, as is the possible relationship of cAMP with other differentiating agents. Several questions concerning the problem of neuronal differentiation are raised for future study. Various studies on the use of hybrid neuronal cell culture are presented, illustrating the advantages of using this type of neuronal culture for future studies. Recent studies on certain aspects of neuronal development, such as morphogenesis, neuronal interaction, cell recognition, neuronal death, and neuronal plasticity are also discussed. In short, this book contains fundamental and recent information on the regulation of differentiation of nerve cells and should be useful as a reference book to neurobiologists, pharmacologists, and cell and cancer biologists.

Kedar N. Prasad

*Denver*

# CONTENTS

## 3. Role of Cyclic Nucleotides in Regulation of Differentiation of Nerve Cells

## 4. Role of Agents Other than cAMP in the Regulation of Differentiation of Nerve Cells

# DEFINITION AND METHODOLOGIES

## INTRODUCTION

The study of the induction and regulation of cell differentiation may hold the key to many current health problems, including oncogenesis and neurological diseases. There are two aspects of neural differentiation: (1) induction mechanisms and (2) regulatory mechanisms. The differentiation of nerve cells from presumptive epidermis involves the induction of many differentiated functions unique to nerve cells. The process of induction involves extensive modification of gene expression, i.e., activation and deactiviation of certain genes, as well as precise patterns of growth and organization. Regulatory mechanisms may be important in maximizing the expression of all differentiated functions in the fully mature neuron, and may thereby facilitate the structural organization of the cell. The level of some or all differentiated cellular functions may be regulated by mechanisms distinct in part from those involved in induction. The understanding of both induction and regulatory mechanisms is important. Although there are several experimental models with which to study regulatory mechanisms, none exists for the study of induction mechanisms in mammalian systems. Various cell culture systems undoubtedly provide relatively simple experimental models with which to study certain aspects of neural differentiation that are preferable to the study of intact organisms. However, the cells that compose these systems are already differentiated; therefore, one can only increase or decrease the expression of various differentiated functions through the use of external agents. Using currently available model systems, it has been possible to identify many molecules important in regulating the level of individual differentiated functions in mammalian nerve cells. Using primarily amphibian gastrulas, it has been possible to identify certain molecules that may be involved in inducing neural tissue in ectoblasts. Although genetic control is necessary for the differentiation of nerve cells, other (epigenetic) factors are required for subsequent normal development of neurons. Among these factors, intercellular communications, cellular interactions, and maternal and fetal circulation are the most important during prenatal development of nerve cells. After birth, however, the infant's own internal and external environments play a vital role in regulating the expression of final stages in nerve cell differentiation. For example, sensory deprivation has catastrophic effects on the development of nerve cells in newborn mammals. The problem of cell differentiation in general and nerve cell differentiation in particular has been discussed in a number of symposia and monographs.[1-12]

## DEFINITION OF DIFFERENTIATION

Differentiation is a process by which the descendants of a single cell, the fertilized egg, come to differ from one another and from the parent cell. Such cells, by the continuous process of differentiation, eventually form tissues and organs performing specialized functions. The differentiation of neural tissue involves the induction, migration, organization, and formation of many anatomically and functionally distinct types of neurons. These neurons form precise nerve circuits between each other and between themselves and nonneural target tissues such as muscle.

## CRITERIA OF DIFFERENTIATION OF NERVE CELLS

Those morphological and biochemical properties unique to nerve cells can be referred to as differentiated functions of nerve cells. These include formation of neurites; formation of synapses; formation of precise neural circuits; uptake, storage, and release of neurotransmitters; high activity of certain enzymes (tyrosine hydroxylase, choline acetyltransferase, dopamine $\beta$-hydroxylase, monoamine oxidase, and acetylcholinesterase) involved in the metabolism of neurotransmitters; increase in size of cell soma and nucleus, associated with a rise in total RNA and protein contents; electrically excitable membrane; and high activities of adenylate cyclase and adenosine 3',5'-cyclic monophosphate (cAMP) phosphodiesterase, associated with high levels of cAMP. Some functions, however, are associated with, but not unique to, differentiated neurons. These include inhibition of cell division and diploid DNA content per cell.

## AVAILABLE MODELS WITH WHICH TO STUDY THE PROBLEM OF DIFFERENTIATION

### A Model for the Study of Induction Mechanisms

Many model systems are available for the study of regulatory mechanisms of differentiation in mammalian nerve cells; in contrast, the amphibian gastrula system is the only model widely used to define mechanisms of neural induction. Explants of presumptive epidermis of three species of amphibians (*Xenopus laevis, Pleurodeles waltili,* and *Siredon mexicanum*) are excised as soon as the dorsal lip becomes visible. The explants are incubated as described by Niu and Twitty[13] and as modified by Barth and Barth.[14] During excision and preparation of cell aggregates (small explants), the medium (10% Ringer's solution) contains 0.077–0.088 M NaCl. However, during subsequent treatment with an inductor, the concentration of NaCl is reduced to 0.044 M. Lyophilized calf serum (0.1%) is added instead of serum globulin. The concentration of EDTA was reduced to 10 mg/100 ml of calcium- and magnesium-free modified Niu–Twitty solution. The agar substrate was omitted during excision healing and treatment of cell aggregates (small explants).

The medium used by Niu and Twitty contains the following components[13]:

Solution A (500 ml)
NaCl—3400 mg
KCl—50 mg
Ca(NO$_3$)4H$_2$O—80 mg
MgSO$_4$—100 mg

Solution B (250 ml)
Na$_2$HPO$_4$—110 mg
KH$_2$PO$_4$—20 mg

Solution C (250 ml)
NaHCO$_3$—200 mg

The solution of Barth and Barth[14] is composed of the following:

Solution A
NaCl—5.150 g
KCl—0.075 g
$MgSO_4 7H_2O$—0.204 g
$Ca(NO_3)_2 H_2O$—0.062 g
$CaCl_2{}^3 2H_2O$—0.0660 g
$H_2O$—500 ml

Solution B
$NaHCO_3$—0.200 g
$H_2O$—250 ml

Solution C
$Na_2HPO_4$—0.0300 g
$KH_2PO_4$—0.0375 g
$H_2O$—250 ml
Serum globulin, 0.1%

The routine practice used by Barth and Barth[14] involves the addition of lyophilized calf serum to solution B, the bicarbonate component of the standard solution. After standing for 5 min, the solution is brought to the simmering point three times over a Bunsen burner flame. This solution is then cooled and added to the "A" and "C" components of the standard solution.

*Procedure for Preparing Cell Aggregates*

Cell aggregates are prepared from amphibian explants as follows,[14] using presumptive epidermis as an example. Jelly is removed down to the vitelline membrane in 10% Ringer's solution, using a watchmaker's forceps. All operations are carried out under sterile conditions. Vitelline membranes are removed with a watchmaker's forceps from a group of eight gastrulae and placed in the standard salt solution containing calf serum. Membrane removal is accomplished by deliberate puncture, either through the yolk or just above the dorsal lip, in order to avoid injury to the presumptive epidermal region. Demembranated embryos are transferred by a wide-mouth pipette to a dish containing fresh standard solution, where presumptive epidermis is excised with a glass needle and hairloop. In the absence of a precise map of the gastrula of *Rana pipiens*, the development of the donor gastrula, from which presumptive epidermis was taken, was observed. The opening into the blastocoel healed rapidly at room temperature, and normal-appearing late neurulae developed within 20 hr. The small wound was located ventrally, about midway along the anteroposterior axis of the embryo, to mark the site of explantation of presumptive epidermis.

Presumptive epidermal explants are passed through the calcium- and magnesium-free EDTA solution in another petri dish for 3–3.5 min. The EDTA treatment is sufficient to loosen the outer, relatively impermeable, pigmented coat cell layer from the underlying presumptive epidermal cell layer. The outer layer is peeled off and discarded. The brief EDTA treatment does not dissociate the inner cell layer into individual cells.

The inner layer is then teased, by means of two hair loops, into 25 aggregates composed of about 125 cells each. The term "aggregates" is used to distinguish these small groups of cells from the large presumptive epidermis explants from which they were made. The small groups of cells (aggregates) should not be confused with "reaggregates," a term used by a number of other workers who utilize protease—EDTA treatments to totally dissociate tissues into individual cells, followed by reaggregation of these cells.

From 8 explants an approximate total of 200 aggregates can be prepared within 30 min from the time of vitelline membrane removal. After heating for 10–15 min, aggregates are transferred, by means of a Spemann pipette, to desired experimental and control solutions. After 3–4 days at 20–22°, the cultures are examined with a compound binocular microscope for attachment and early differentiation. Subsequent observations are made as often as needed in order to follow the progress of differentiation. By enclosing the microscope in plastic sheeting, contamination during observation is minimized. Features characteristic of differentiation are, in most cases, expressed by 5–7 days. To obtain a permanent record, cells can be cultured on cover glasses on the floor of culture dishes.

## Models for the Study of Regulatory Mechanisms

The cells of mammalian neural tissues divide, differentiate, migrate, and mature in a highly ordered pattern. Each neuronal function is expressed at a precise time for a well-defined purpose. A defect in the regulation of any of the above steps can result in abnormal neural tissue. Depending upon the type of lesion, the stage of development during which the lesion occurs, and the type of cell (neuronal vs. glial) involved, such an abnormality could be expressed as malignancy or as one of a variety of neurological disorders. Therefore, an understanding of the regulation of growth rates, differentiation, migration, maturation, and the expression of individual differentiated functions of nerve cells would be helpful in developing new approaches to the therapy of neurological diseases. Obviously, the above problems cannot be systematically investigated in a complex experimental system such as that found *in vivo*. During the last 70 years, attempts have been made to investigate various aspects of neurobiology by growing neural tissues in a well-defined chemical environment outside the body under *in vitro* conditions. Using such growth conditions, remarkable progress has been made with respect to elucidating certain aspects of regulation of expression of differentiated functions.

Six major types of *in vitro* techniques have been used to investigate the regulation of differentiated functions and malignancy in nerve cells:

1. Explant culture.
2. Aggregate culture.
3. Dissociated cell culture.
4. Monolayer cell culture.
5. Suspension cell culture.
6. Culture of whole ganglia.

None of the above experimental systems is suitable for the study of all aspects of neurobiology; rather, each system offers unique advantages and disadvantages for exploring a given aspect of nerve cell function. The purpose of this chapter is to describe the techniques involved in each type of culture system, and to discuss the usefulness and limitations of each system as applied to the study of regulation of growth rate, malignancy, and differentiation of nerve cells.

*Explant Culture*

In 1907, Harrison pioneered the use of neural explant culture.[15] He explanted a small portion of neural tube from a frog embryo and demonstrated the subsequent development of an axis cylinder from perikaryon over clotted lymph. This technique has since been used extensively to provide better understanding of such problems as cellular morphology and movement, neurite outgrowth, neurophysiological properties, neurosecretion, effects of drugs and nerve growth factors, ultrastructure of neurons in long-term culture, maintenance and formation of new synapses, myelination, axoplasmic flow, and experimental allergic encephalomyelitis. Neural tissue thus explanted can fully achieve cytotopic, organotypic, and functional maturation *in vitro*.[16-21] Various investigators have employed a variety of explant culture techniques; these are described below.

Explant culture techniques for the growth of neural tissues are now well refined. The most commonly utilized tissues for such studies include cerebellum, cerebrum, dorsal root ganglia, and autonomic ganglia obtained from neonatal rats, kittens, or chick embryos. The most commonly used procedure for explant culture is the double coverslip modification[22] of the Maximow assembly. Small pieces of neonatal tissue (approximately 1 mm square) are placed on collagen-coated[23] coverslips, and this round coverslip is fixed by capillary action to a larger, square coverslip. The explant is fed with a drop or two of nutrient medium containing a balanced salt solution, serum, embryo extract, and glucose. When the explants are firmly in place and fed properly, the assembly is completed by placing the double coverslip over a depression in a glass slide. The chamber is then sealed with paraffin and incubated at 35°. In order to feed such cultures or to modify their nutrient medium, the inner round coverslip is removed, washed, placed on a new coverslip, refilled with the appropriate medium, and sealed onto a new depression slide.

The nutrient medium used in explant culture varies from one investigator to another and from one tissue to another. For 18-day-old mouse hippocampus or neocortex,[24] the nutrient medium usually contains human placental serum (33%), Eagle's minimum essential medium (MEM) (53%), chick embryo extract (10%), glucose (600 mg/100 ml), and achromycon with ascorbic acid (1.2 $\mu$g/ml). Explants[25] of rat spinal cord (15.5-day-old) and rat superior cervical ganglion neurons (19- to 21-day-old) are generally cultured in 88% MEM, 2% 9-day-old chick embryo extract, 10% human placental serum, 600 mg/100 ml glucose, 20 U/ml nerve growth factor (NGF), $10^{-5}$ M 2'-deoxy-5-fluorouridine, $10^{-5}$ M uridine, and 15 mM KCl. Increased potassium levels favor neuronal survival in culture.[26] The addition of 2'-deoxy-5-fluorouridine prevents the growth of dividing nonneural elements and thereby allows one to observe neuronal neurite growth.

A different explant technique has been used for culture of chick embryo sympathetic ganglia.[27] The chick embryo, at 13–15 days of incubation, is suitable for culturing explants of sympathetic ganglia. Embryos are removed aseptically from their shells and placed in a sterile petri dish. After discarding any embryos showing developmental abnormalities, the lumbar sympathetic chains are removed under a dissecting microscope fitted with a 20× objective. The first ganglion of the chain and its contralateral partner are used.

The culture system used is a modification of that of Basrur et al.[28] using a Leighton tube. In brief, aseptically obtained tissue fragments are rinsed in Hank's balanced salt solution and placed in a petri dish containing a few drops of growth medium. The tissue is chopped into fine pieces of 1 mm$^3$ or less, and six or eight such pieces are then drawn into a curved Pasteur pipette along with a small amount of medium. These tissue fragments are distributed over the surface of an

11 × 35 mm coverslip on the floor of a Leighton tube, and the excess medium is drawn off. A second coverslip is placed over the explant in such a way that its edge overlaps that of the lower coverslip by a few millimeters. Each Leighton tube with its explants is then placed with the flat surface upwards, on a slightly sloped rack, and 2 ml of growth medium is then pipetted over its round surface. Five or six such tubes, sealed with siliconized rubber stoppers, are placed on their flat surfaces in the rack and incubated at 38°. The only difference between this and the original technique is the use of a single rather than a double coverslip system. The ganglia are pipetted on the glass coverslip and covered with 1.5 ml culture medium. The Leighton tube is then aerated with 95% air–5% $CO_2$ and incubated in an oven at 37°. The coverslips are prepared by treating them for 1 hr in 2 N HCl, rinsing thoroughly, washing with water, and, finally, rinsing with distilled water. Coating the coverslips with collagen is unnecessary for the short culture period of 3–4 days. The culture medium contains 90% medium 199 supplemented with 500 mg/100 ml D-glucose and 10% fetal calf serum. NGF is added in varying concentrations (1–20 U/ml). Embryo extract is not used. This method is generally 100% effective. It is simple and rapid, since 10 ganglia can be grown on a single coverslip and 20 tubes can be set up in a morning's work. Oxygen tensions in the medium under these conditions are approximately threefold greater than those present *in vivo*, but the $CO_2$ tension and pH values are within the physiological range.

Another method can be used for culturing explants of sensory ganglia.[29] Sensory ganglia are dissected out from 7- to 10-day-old chick embryos and collected in physiological solution. They are then transferred into 0.5% trypsin solution and incubated at 37° for 20 min. After careful washing in several changes of physiological solution, the ganglia are gently dissociated by suction with a micropipette and dispersed in 0.5 ml of MEM, supplemented with 10% horse serum. A homogenous suspension of nerve and supporting cells is thus prepared and four drops of this suspension are placed in plastic dishes containing nutrient medium. These dishes are maintained at 4% $CO_2$ in air. Cells of sympathetic ganglia from 9- to 14-day-old chick embryo are prepared in a similar manner, except that trypsin exposure time is longer (35 min). The volume of medium in each dish is 1.5 ml. NGF is added daily in a final concentration of 0.05 $\mu$g/ml.

*Advantages.* (1) Neural tissue under the above conditions can achieve full cytotopic, organotypic, and functional maturation *in vitro*. (2) Since these explants are maintained in a well-defined chemical environment, the effects of pharmacological agents, hormones, and nerve growth factor can be evaluated without indirect effects of other agents present *in vivo*. (3) Environmental variables can be manipulated in either numerical or temporal combinations that are not possible *in vivo*. (4) Precise cell-to-cell contact is maintained, similar to that found *in vivo*. The cellular architecture and assembly of explant cultures closely resemble *in vivo* conditions.

*Disadvantages*. There are several disadvantages of explant culture. (1) Although explants are very small, centrally located cells frequently receive fewer nutrients and may thus become necrotic. (2) The expression of biochemical properties cannot be studied in separate neuronal and nonneuronal populations. (3) Cells within the explants cannot be visualized, and this limits the study of morphology of intracellular components. (4) Some cells within explants die, and many others migrate out along the attachment surface until the explant becomes flattened and fixed. Neurons are generally passive but may be pulled around other cells. This causes a dynamic change in the explant culture, which is unique to *in vitro* growth conditions. (5) Neuron populations in explants generally do not divide, whereas glial cells do; therefore, the regulation of growth rate of nerve cells cannot be studied. (6) The sample size for biochemical analysis is small.

### Aggregate Culture

The aggregate culture technique allows dissociated cells to maintain cell-to-cell contacts and therefore produces histotypic patterns characteristic of the original tissue. This technique has been used extensively by Moscona[30,31] for various neurobiological studies. Many biochemical and morphological features of differentiation are expressed in aggregate culture in a fashion similar to that observed *in vivo*.[30-42] The tendency to aggregate is a property of all embryonic cells; and, in general, the more undifferentiated the tissue, the better the aggregation. Several factors influence cellular aggregation. These include tissue of origin, state of cellular differentiation, rotation speed *in vitro*, temperature, calcium concentration, serum proteins, and cell-specific aggregation factors.[34] Aggregation is enhanced by gently rotating a suspension of trypsin-dissociated cells in Erlenmeyer flasks on a gyrotary shaker. A speed is selected (about 70 rpm) such that the cells are brought into a vortex, thereby greatly increasing the number of collisions between cells. At higher speeds, the shearing forces increase; therefore, the aggregate size decreases, but the number of aggregates per flask increases. The volume and composition of the cell culture medium are also important variables. Aggregation also directly depends upon the temperature. Maximal aggregation occurs at 37–38°. Lower temperatures reduce the aggregate size, and temperatures below 15° completely block aggregation. New protein synthesis is known to be necessary for reaggregation, since puromycin, an inhibitor of protein synthesis, inhibits this phenomenon. The aggregation of cells from a specific tissue decreases with increasing differentiation. Although the technique of reaggregation has been described by several investigators, the methods described below are taken from a recent study by Seeds *et al.*[35]

*Cell Dissociation.* C56BL/6J mice (17 days pregnant) are used as a source of fetuses for cell culture. The entire fetal brain is removed and finely minced. The tissue is dissociated in a solution of 0.25% trypsin (Difco 1:250) in saline (0.138 M NaCl, 5.4 mM KCl, 1.1 mM $Na_2PO_4$, 1.1 mM $KH_2PO_4$, 0.4% glucose, and 0.01% $CaCl_2$), and incubated at 37° for 15 min, with constant rotation. The tissue is dispersed by gentle pipetting (three times). Large and undispersed pieces of tissue are allowed to settle out, and the suspended cells are removed and passed through a nylon screen to complete the dissociation into single cells. The larger pieces are swirled in fresh trypsin solution for an additional 10 min, then pipetted and screened as before. After the addition of fetal serum to final concentration of 10%, the cells are collected by centrifugation at 200 g for 5 min.

*Cell Culture.* Aggregate cultures are prepared from $0.5 \times 10^7$ to $1 \times 10^7$ cells, in 25-ml Erlenmeyer screw-cap flasks containing 3 ml of basal Eagle's medium with 0.4% glucose and 10% fetal calf serum. The flasks are equilibrated with 5% $CO_2$ in air and incubated at 37° with constant rotation (40 rpm). Aggregation is complete within 12–24 hr. These initial aggregates are composed of loosely packed and randomly dispersed cells, and have small cellular extensions. After several days, there is an increase in cellular outgrowth, thus forming a more stable complex. During this time, the cells undergo extensive migration and reaggregation within the aggregate into clusters of similar cell types. The ability of cells derived from different tissues to sort out in aggregates of mixed cell populations has been known for many years.[36,37] Since many cell types are found in brain tissue, and since various brain regions develop at different rates, it is not surprising that cells of specific brain regions segregate themselves from cells of other regions.[38]

The technique of aggregate culture has been used to study biochemical properties of developing chick retina.[39]

*Advantages.* Certain advantages of this culture system are similar to those described above for explant cultures, but there are some additional advantages as well. The relatively large amount of nerve tissue obtained can be used to analyze many biochemical functions associated with cellular differentiation. This system also permits the study of cell-surface properties relevant to histogenesis.

*Disadvantages.* The disadvantages of aggregate culture are similar to those described below for dissociated cell cultures. There may be some additional disadvantages. The individual cells may be altered to varying extents during the dissociation process. Since similar kinds of cells tend to aggregate with themselves, the influence of different cell types, if any, may thus be lost.

### Dissociated Cell Culture

The problems posed by cellular heterogeneity, viability of neurons, and complex interrelationships among various cell types in explant culture necessitated developing a method for isolating neuronal and nonneuronal components of neural tissue in order to study the function of neurons and glial cells separately. Such cell populations could then be used to study the effect of interaction of these cell types *in vitro* on the expression of various cellular functions, which then can be compared with the properties of original undissociated nerve tissue. An elegant technique of isolating individual neuronal and nonneuronal cell types, which involves dissociation, fractionation, and subsequent culture, has been developed by Varon and Raiborn.[40] The following procedure is carried out under sterile conditions and at room temperature: (1) The 11-day-old chick cerebrum is selected as a starting material for the isolation of various cell types, because neural differentiation is practically complete by 12 days of incubation. Following day 12, neuroglia increase; fiber tracts thicken; important metabolic changes, including initiation of sphingomyelin synthesis, take place; and bioelectrical activity appears. (2) The tissue, dissected free of its meningeal membranes, is placed in a nylon bag (202 mesh), which is then immersed in Eagle's basal medium (EBM) (4 ml per cerebrum) and stroked on the outside for a few minutes until it no longer releases cellular material into the external fluid. Following this mechanical dissociation and suspension of tissue, the external medium contains a large number of dispersed cells, various cell aggregates, and minute cell debris. (3) The resultant cell suspension is further filtered on nylon cloth (30 mesh) in order to remove large aggregates, centrifuged for 2 min at 700 rpm, and suspended in an equivalent volume of buffer. (4) 4-ml aliquots of the above suspension are layered in round-bottom polyethylene tubes over 5 ml of a 1:1 mixture of 15% bovine serum albumin (fraction IV) and 20% sucrose, both in EBM. The tubes are centrifuged for 10 min at 700 rpm. The upper phase is discarded, and the interface and underlying bovine serum albumin–sucrose phase ($S_1$) is collected. The pellet ($P_1$) is resuspended in 7 ml of fresh EBM and is layered over bovine serum albumin–sucrose and centrifuged as above. The second $S_2$ and second $P_2$ are obtained. $S_1$ and $S_2$ are pooled, diluted with an equal volume of EBM, and centrifuged for 5 min at 1000 rpm. The sediment is resuspended in 4 ml of EBM (S fraction). The $P_2$ pellet is also suspended in 4 ml of EBM. The $P_2$ and S fractions may be pooled if desired. (5) An equivalent volume of calcium- and magnesium-free balanced salt solution, containing 0.002% trypsin, is added to the solution containing $P_2$ and/or S fractions. After 2- to 3-min incubation, the suspensions are centrifuged for 5 min at 1000 rpm. The resultant pellets are resuspended in small volumes of EBM and are thereafter referred to as $P_2T$ or ST, for trypsin-treated $P_2$ or S fractions. (6) three-milliliter aliquots of $P_2T$, containing 20% calf serum, are seeded in T-15 flasks and incubated at 38° for 24 hr. Medium containing unattached cells is then withdrawn from the flask and centrifuged for 5 min at 700 rpm; the pellet is resuspended in EBM. This fraction is approximately 90% pure and contains primarily A cells. The cells that remain attached in each flask ($P_2TA$) are harvested in 3 ml of $Ca^{2+}$- and $Mg^{2+}$-free balanced salt solution, containing 0.1% trypsin, and centrifuged for 5 min at 1000 rpm. The suspension of this fraction contains two distinct types of cells, B and C, as well as numerous small cells; however, these cells are not readily distinguishable from one another. The $P_2TA$ suspension is seeded on a glass surface containing EBM and 20% calf serum. After 24 hr, only C cells are attached. This pure population of cells is designated CC. It is not yet possible to isolate B cells and small cells in a similarly purified form. Thus, three major cell types can be isolated by the technique described above.

*A Cells.* When grown in monolayer culture, these large round cells develop numerous processes. Most A cells are multipolar, with cell bodies 15 $\mu$m in diameter. Their nuclei are distinct, are usually located eccentrically, and contain two nucleoli. The A cells do not attach to glass even in a serum-supplemented media. Mitotic activity has not been demonstrated in these cells, which probably represent well-differentiated neurons.

*B Cells.* The B cells are small (7–10 $\mu$m in diameter) and dense and have smaller nucleoli than A cells. They also develop processes, but unlike A elements, B cells are usually bipolar. B cell fibers, unlike A cell fibers, do not take up silver stain; however, like A cells, B cells do not attach to glass and do not proliferate. They also probably represent neuronal elements.

*C Cells.* The C cells are similar to B cells, but they behave differently in culture. C cells spread out into very large, extremely thin multiform elements having ragged contours and gross processes. The nuclei are oval and usually centralized. These cells attach to uncoated glass surface only in the presence of serum-supplemented medium. These C cells are nonneural cells.

Cultures prepared as described above develop very rapidly. A and B cells start putting out processes within a few hours of plating; by 4 hr, many of these processes are long and branched. By 24 hr, long fibers extend across the field in complex lattices. These fibers make numerous contacts with each other. They can make intricate arrangements or appear to merge with one another; or they may bundle up into tight, nervelike fascicles. Fiber-to-cell contacts are also frequent between A cells, or between A and B cells. Most A and B cells start dying out after 4–7 days of culture, although a few such elements have been seen to survive without apparent change for over 1 month. By day 4, in most cultures proliferation of C cells becomes the dominant feature, and by day 7 these cells have replaced the other cell populations.

Varon and Raiborn[41] used somewhat different methods to prepare dissociated cell cultures from chick embryo sympathetic ganglia. The lumbar halves (10 ganglia each) of the paravertebral sympathetic chains are dissected from 7- to 15-day-old chick embryos and are collected in either Hank's balanced salt solution (BSS) or $Ca^{2+}$- or $Mg^{2+}$-free saline (CMF; 1 liter of solution containing NaCl, 8 g; KCl, 0.2 g; $Na_2HPO_4 \cdot 2H_2O$, 0.05 g; $NaHCO_3$, 2.2 g; and glucose, 6 g) at pH 7.5. A ratio of 1 ml of collecting solution per four half-chains (40 ganglia) is used. Tissues from as many as 10 embryos are pooled, with a maximal collection time of 30 min. At the end of the collection step, the collection fluid is aspirated off and replaced with an equal volume of BSS, containing 0.25% trypsin; this suspension is incubated at 37° for 30 min. After a brief contrifugation, the trypsin solution is replaced with an equal volume of low-bicarbonate culture medium (LMS) consisting of EBM supplemented with glucose (increased to 6 g/liter), penicillin (200,000 U/liter), fetal calf serum (5%), and 7 S NGF (100 BU/ml). Full dispersion of tissue in LMS fluid is achieved by 10 aspirations through a Pasteur pipette (approximate tip diameter 500 µm). The dissociated cell suspension (harvest) is differentially counted for neuronal and nonneuronal elements in a hematocytometer chamber under a phase contrast microscope. The neuronal populations could be distinguished by their size and shape, namely, larger (20 $\mu$m or greater in diameter) polygonal elements and smaller (about 8 $\mu$m in diameter) round ones. The nonneuronal cells are easily distinguished from the neurons by their considerably smaller size (5 $\mu$m or less in diameter), their irregular contours, and the presence of cytoplasmic granulations.

For culture purposes, the harvest suspension is diluted approximately 8-fold, to 50,000 neurons/ml, using high-bicarbonate culture medium (HMS) having the same composition as the LMS, except for a 10-fold increase in bicarbonate level, to 2.2 g/liter. Two-milliliter aliquots (100,000 neurons, plus an approximately equal number of accompanying nonneuronal cells) are seeded into 35-mm plastic Falcon dishes, precoated with reconstituted rat-tail collagen, and the dishes incubated at 37° in a water-saturated, 5% $CO_2$–95% air mixture. For long-term culture, the medium was changed every other day.

The collection of ganglia directly into $Ca^{2+}$- and $Mg^{2+}$-free saline, rather than into BSS or low-bicarbonate culture medium, results in a considerably higher yield of both neuronal and nonneuronal cells, with no noticeable impairment of their subsequent viability in culture.[41] The relative advantage of CMF collection is considerably greater for younger ganglia (almost 3-fold) than for older ones (about 1.5-fold). The duration of the tissue exposure to either CMF or BSS in the collection step can be varied from 2 to 60 min with no noticeable influence on the final dissociation yield. Lowering the trypsin concentration or shortening the time of treatment progressively reduces final cell yields, but increasing either or both parameters does not provide significant improvement. Omission of 7 S NGF from the dispersion medium does not result in lower yields. Prolonged exposure to alkaline medium (pH 9) at any stage of the dissociation procedure decreases the final cell yield. The ratio of medium to tissue is an important factor, as ratios lower than 25 $\mu$l per ganglion, or greater than 75 $\mu$l per ganglion resulted in significantly poorer yields of both neuronal and nonneuronal elements.

Booher and Sensenbrenner[43] have compared the growth and cultivation of dissociated neurons and glial cells from embryonic chick, rat, and human brain grown in flask cultures with those grown in flasks coated with reconstituted rat-tail collagen. Various ages of embryonic brain tissue from each species were studied (5-,7-,8-,11-,14-, and 18-day-old chick embryos; 7-,9-, and 16-day-old rat embryos; and 13- and 19-week-old human embryos). Prior to dissociation, meningeal and connective tissues are removed from the brain with the aid of a stereodissecting microscope, in order to assure that only cerebral and subcortical tissue cells are used for culture. Two standard techniques may be employed for the dissociation of cerebral tissue: trypsin (0.25%) dissociation or the sieve technique, in which brain tissue is gently passed through a nylon mesh (48-$\mu$m pore) into a small reservoir of nutrient medium. The innoculum of cell suspension is between $2 \times 10^5$ and $2.5 \times 10^5$ cells/ml in all culture preparations. The nutrient medium consists of EBM supplemented with 20% fetal bovine serum, 20% (v/v) whole egg ultrafiltrate, 600 mg% (w/v) glucose, penicillin (100 U/ml), and streptomycin (100 $\mu$g/ml). The pH of the medium is maintained between 6.9 and 7.3. During the growth period, the medium is changed periodically. The cell cultures are incubated at 37° in an atmosphere of 95% oxygen and 5% $CO_2$. Relative humidity is maintained between 80 and 90%. Subcultures are made by trypsinizing the cells and reseeding them into either untreated or collagen-coated flasks, at one-half of the original cell concentration. The basic difference between cultures grown on the collagen-coated surface and those grown on plastic alone is evident during the first week of culture. Cells in collagen-coated flasks remain well isolated, as opposed to the clumping of cells that is observed in cultures grown in untreated plastic. Following the establishment of a monolayer of mesenchymal cells, upon which neuronal elements appear to develop and differentiate, the morphological aspects of the two types of culture are nearly identical.

With respect to neuronal survival and maturation, the best result can be obtained with tissues from 8-day-old chick and 7-day-old rat embryo. Cultures made from older chick and rat brain result in a high yield of mesenchymal cells, along with numerous glial-like elements. Both 13- and 19-week-old human fetal dissociated brain cells can be grown and differentiated into mature neurons on collagen and on plastic. As with tissues derived from rat and chick brain, these human brain specimens are obtained at a time near midgestation, prior to the onset of complete morphological neuronal differentiation in the human brain.

Several other methods of isolating neural and nonneural cells have been used. The free-hand dissection of individual neurons from frozen section,[42] or of individual neurons and glial clumps from fresh sections,[44] has been carried out. Larger numbers of individual neurons have been isolated by mechanical disruption of tissue followed by filtration of the suspension through nylon sieves.[45] Preparation of dissociated neuronal and glial cells on a larger scale[46-48] has been accomplished by following the procedure for other tissue.[49] Other have tried to fractionate cell suspensions by differential centrifugation in sucrose,[49,50] sucrose-serum mixtures,[51,55] Ficoll, and acetone–glycerol–water.[53]

Another study[55] has reported the isolation of viable neurons from embryonic chick spinal ganglia by centrifugation through albumin gradients. Thirty to 40 dorsal root ganglia are removed from 7-day-old chick embryos and are immersed in a petri dish for 5 min at 37° with 5 ml of 0.25% trypsin in BSS. The ganglia are gently pipetted back and forth for 10 min with a small-bore Pasteur pipette, in order to remove the capsule and intact nonneuronal tissue. The cells are then incubated in the trypsin solution for 15 min at 37°. The ganglia are dissociated by continuous pipetting and incubating for approximately 1.5 hr. The mixed cell suspension is then centrifuged at 100 g for 2 min; and 1 ml of medium 199, supplemented with 700 mg/100 ml glucose, is then added to the pellet and the suspension is prepared.

The neuronal fraction is separated from other cells in suspension by ultracentrifugation in a Spinco Model L (SW 39 rotor), using bovine serum albumin gradients in the following specific gravities: 1.100, 1.079, and 1.065. Layering of the gradient fractions is done at 4° and the tubes are kept cold at all times. One milliliter of each albumin fraction is added to a different sterile 5-ml centrifuge tube, with the highest density on the bottom. The tubes are centrifuged at 30,000 rpm for 1 hr at 4°, and the enriched neuronal fraction is collected from the 1.079 and 1.100 interface. A culture is prepared by mixing 1 ml of the cell suspension with 1 ml of medium 199, containing 400 mg/100 ml glucose, and 1 ml of fresh rooster plasma and nerve growth factor, and a portion of the mixture (1 ml) is placed in the rose chambers. Within 48 hr, fiber outgrowths from the neurons attain dimensions 12 times the length of the original cell soma.

Viable neurons can also be isolated from mechanically dissociated cultures by centrifugation through albumin gradients. The advantages and disadvantages of this technique are identical to those described in the Dissociated Cell Culture section. It is impossible to state whether the technique of Varon and Raiborn is better than this until a comparative study on both techniques is made.

*Advantages.* This technique provides an opportunity to compare the behavior and biochemical properties of nerve and nonnerve cells separately and in conjunction with one or more cell types. This may clarify which neural functions depend upon perfect tissue organization and cell-to-cell contact for maximal expression.

*Disadvantages.* One cannot study the induction and regulation of expression of differentiated functions, since these cells are already fully differentiated. One cannot study growth-rate regulation in nerve cells, because nerve cells obtained after dissociation do not divide in culture. The following disadvantages are characteristic of *in vitro* systems and are not necessarily unique to dissociation techniques: (1) The original tissue organization no longer exists; (2) cell–cell contacts are lost; (3) intercellular distances are increased to a degree depending on the volume of fluid in which the cells are dispersed; (4) numerical balance among different cell types is changed; (5) individual cells may be altered to varying extents, depending on the type and source of tissue and the procedure used; and finally, (6) although trypan blue exclusion has been used to estimate the viability of isolated cells, this is a highly unreliable method.

*Monolayer Cell Culture*

None of the techniques previously described is suitable for the study of regulation of growth rate, differentiation, and gene expression in homogeneous populations of nerve cells. The availability of monolayer cell culture provides a unique opportunity to study these problems. Unfortunately, no such culture of normal dividing nerve cells is available. However, monolayer cultures of nerve cells derived from tumors are now available. Monolayer cultures of mouse and human neuroblastoma are are now used in several laboratories to investigate the problems of regulation of growth rate, differentiation, malignancy, and various other neurobiological problems. Many responses of neuroblastoma cells to various agents are similar to those observed with explants of embryonic nervous tissue.[56] We will describe separately the techniques for culturing mouse and human neuroblastoma.

*Mouse Neuroblastoma.* The original mouse Cl300 tumor arose spontaneously in the body cavity of a mouse (strain A/J) in 1940, and has since been maintained by subcutaneous transplantation in A/J mice at the Jackson Laboratory, Bar Harbor, Maine. This tumor was initially diagnosed as a round-cell tumor, possibly neuroblastoma.[57] However, its neuronal character was not confirmed until 1969, when two groups of investigators[58,59] independently established the cell line in monolayer culture. Schubert *et al.*[59] used the following technique. The solid tumor is adapted to tissue culture conditions by dispersing the cells in modified Eagle's medium containing 20% fetal calf serum. The cell cultures are maintained at 37° in an 85% air and 15% $CO_2$ incubator. Cells are cloned twice by spreading dilute cell suspensions on solid agar (0.5% agar, 4.5 mg/ml trypticase soy broth in Eagle's modified medium plus 20% fetal calf serum) and picking visible colonies with a platinum loop after a 2-week incubation.

Augusti-Tocco and Sato[58] used a different technique but obtained similar results. The technique of alternate passage animal culture[60] was followed to adapt the neuroblastoma cells to monolayer growth conditions. The tissue is dissociated by treatment with Viokase (0.25%) solution, and single cells are plated in Falcon plastic flasks, pretreated with 5% gelatin solution. Ham's F-10 medium, supplemented with 15% horse serum and 2.5% fetal calf serum, is used. Clonal lines are isolated following the single-cell plating technique described by Puck et al.[61] from cultures at the second or third passage in vitro.

In our laboratory,[62] we follow the method described by R. Klebe of Yale University. Solid tumor, free of necrotic tissues, is dissected out. The solid tumor of 1 cm × 1 cm has very few or no necrotic cells. Solid-tumor tissue is washed three times in $Ca^{2+}$-free MEM and placed on an organ culture grid immersed in $Ca^{2+}$-free MEM. The tissue is grated by organ culture grid. Clumps of tissue float away into the medium. The remaining fibrous tissue is excluded from further treatment. Clumps of tissue are washed twice with $Ca^{2+}$-free MEM and incubated in the presence of Viokase solution (0.25% in $Ca^{2+}$-free medium) for 20 min. A single-cell suspension is prepared and cells are washed twice with F-12 medium. Cells are plated in Falcon plastic flasks containing F-12 medium with 10% agammaglobulin newborn calf serum, penicillin (100 U/ml), and streptomycin (100 $\mu$g/ml) and are maintained at 36° in a humidified atmosphere of 5% $CO_2$ in air. Clones are isolated as follows.[63] Two hundred cells are placed in Falcon plastic dishes (60 mm) and are allowed to form clones. A stainless-steel ring coated with nontoxic Vaseline is placed on a well-isolated clone and the ring is half-filled with 0.25% Viokase. After a 5-min incubation, cells are transferred to another dish. The procedure is repeated at least twice before a clone is selected for an experiment.

In spite of variations in techniques used for monolayer culture by different laboratories, the morphology and biochemical features of neuroblastoma cells are expressed in a similar fashion. In our experience, we have found the following factors to be important for healthy growth of neuroblastoma cells in culture. (1) Cells should not be allowed to reach confluency before splitting, because confluent cells do not grow well after replating. They may, however, recover eventually. (2) The daily changing of medium, commencing 2 days after plating ($0.25 \times 10^6$ per 75-cm² flask), is essential, because neuroblastoma cells produce lactic acid at a relatively high rate. (3) The medium, Viokase, and centrifuge tubes should be warmed to 37°, and should be used immediately for splitting. In addition, the medium should be prewarmed in the flask in which new cells will be plated. The above steps were found to be necessary to attain good growth conditions. The cells can be maintained in good condition at both 36° and 37°.

*Human Neuroblastoma Cells.* Several laboratories now maintain monolayer cultures of human neuroblastoma. Like mouse neuroblastoma cells, human neuroblastoma cells can be adapted to monolayer culture conditions by using different methods and different growth conditions, but the cells behave similarly under all these conditions. Tumilowicz *et al.*,[64] have used the following procedures. Neuroblastoma tissue was obtained from an abdominal mass of a 13-month-old Caucasian boy during exploratory surgery. This tumor mass has rare areas of organoid differentiation. Cells (IMR-32) are explanted by mincing tissue into fine fragments with a sharp blade and placing the resulting suspension in milk dilution bottles containing 10 ml of growth medium. To facilitate the attachment of fragments, the bottles are inverted so that adhering fragments are not in contact with medium introduced on the opposite side. After 1 hr of incubation at 37°, when fragments have become more firmly adherent, the bottle is turned so that the medium covers the fragments. When abundant outgrowth occurs, the remaining fragments are removed with 0.25% trypsin in Hank's balanced salt solution (trypsin and medium components, except antibiotics). Cells are grown in Eagle's minimum essential medium plus nonessential amino acids in BSS and 30% heat-inactivated (30 min at 58°) fetal bovine serum containing penicillin (100 U/ml) and streptomycin (100 $\mu$g/ml). Beginning with the 37th subcultivation, growth medium consists of 80% medium 199 and 20% fetal bovine serum (not heat-inactivated), containing antibiotic concentrations as described above.

Growth medium is usually replaced twice weekly. Temperatures for cultivation range between 35.5° and 37°. Cells are detached from glass for subcultivation with trypsin when the cell layer becomes confluent.

The doubling time of IMR-32 is about 48 hr, and the chromosome number varies from 46 to 48.[64] This cell line is available from the American Tissue Culture Collection. We have used this cell line for several studies and have made the following technical observations: (1) Cells are extremely sensitive to acidic pH; therefore, the medium must be changed frequently; (2) cells are replated at the time when the culture has reached subconfluent stage; and (3) growth rate is better and cells are healthier when the initial number of cells plated in Falcon flasks is at least $0.25 \times 10^6$.

Biedler et al.[65] have used the following procedures. A bone marrow sample obtained from a patient with neuroblastoma is cultured in Eagle's minimum essential medium supplemented with 20% fetal bovine serum, penicillin (100 U/ml), streptomycin (100 $\mu$g/ml), and fungizone (2.5 $\mu$g/ml) in plastic flasks. Erythrocytes are washed off after 2 days, at which point attachment and some growth of cells can be noted. Cultures are transferred three to four times at irregular intervals. After 3 months, SK-N-SH cells are routinely transferred every 2 weeks with replacement medium on days 5, 9, and 12.

The monolayer culture is also established from neuroblastoma tumor tissue obtained from different patients. A small piece of tumor tissue is minced in the medium described above, and tumor pieces are treated with 0.125% trypsin and 0.02% EDTA in calcium- and magnesium-free phosphate-buffered salt solution in order to obtain a cell suspension. Clumps of cells become attached to plastic culture flasks after 2 to 3 days, and by about the 2nd month, SK-N-MC cells could be transferred at weekly intervals, with medium replacement occurring on day 5. In plastic flasks, cells become attached after approximately 24 hr, while in glass vessels, attachment occurs only after 2 days. Both cell lines are routinely maintained in MEM containing 15% fetal bovine serum, penicillin, streptomycin, and nonessential amino acids (Eagle's formulation). For culture transfer, cells are exposed to the EDTA–trypsin solution for 3–5 min.

The chromosome number per cell varies from 44 to 50 with a modal chromosome number of 46.[65] These cell lines are available only in Dr. Biedler's laboratory.

Goldstein et al.[66] have maintained the NJB cell line in their laboratory since 1964, using the following technique. Small fragments of neuroblastoma tissue are explanted under perforated cellophane in fluid media, or in chicken plasma clot cultures on perforated cellophane, in D-35 Carrel flasks. Cultures are also prepared on coverslips in plasma clots in 50-mm petri dishes and gassed with 5% $CO_2$ and 95% $O_2$. The fluid medium consists of either 60% medium 199 or a modified MEM, containing higher concentrations of the essential amino acids, nonessential amino acids, and either 50% human serum or 40% calf serum. Glucose is added at a concentration of 5 mg/ml in some cultures. Penicillin is initially added at a concentration of 200 U/ml.

*Rat Central Nervous System.* A clonal line of rat central nervous system (CNS) tumors was developed by Schubert et al.[67] Neoplasms were induced transplacentally with nitrosoethylurea. BD1X rats were injected 15 days after conception with 4 mg of nitrosoethylurea per 100 g body weight, a treatment which reduced litter size by 50%. Between 4 and 10 months after birth, approximately half the offspring showed symptoms of extreme nervous system disorders and were examined for tumors. In 93% of the animals, tumors were found in the CNS; the other animals had tumors elsewhere. At least 4% of the tumor mass yielded permanent nerve cell lines; the remaining yielded glial cell lines. Excised tumors were finely minced in modified MEM containing 20% fetal calf serum. The cell suspension (including pieces of tissue) was diluted and $10^4$–$10^6$ cells were plated in 60-mm Falcon tissue culture dishes. Each dish was examined periodically for cell proliferation, and since areas of extensive growth were found, cells were isolated by cloning rings and transferred to another dish. Since the initial explants contained cells of widely different morphologies, the most complex cells types were selected. When these cells reached confluency, they were either cloned or passaged for further study. All cell lines were diploid.

*Advantages.* There are several unique advantages to monolayer culture of nerve cells. It would be ideal if such a culture could be established for normal dividing neuroblasts; however, all monolayer cultures of nerve cells have at present been derived from tumors. Nevertheless, such a culture system provides, for the first time, an opportunity to study the following: (1) the regulation of growth rate, differentiation, and maturation in a homogenous population of nerve cells; (2) the expression of individual gene products as measured in dividing nerve cells; (3) the toxicity of pharmacological agents on nerve cells; (4) the mechanism of action of neurotransmitters on the molecular level; (5) the effects of a given agent on separate populations of nerve and glial cells; and finally, (6) recent successes in synapse formation between hybrid neuroblastoma cells (neuroblastoma × glial) and muscle cells[68] provide a unique opportunity to study the regulation of synapse formation.

*Disadvantages.* (1) The monolayer cell culture is completely devoid of *in vivo* tissue architecture, and therefore the information obtained from such cultures may not be pertinent to *in vivo* conditions. (2) The expression of certain genes is lost, while the expression of others is increased. For example, the activity of glucose-6-phosphate dehydrogenase *in vivo* is less than that observed in culture; the reverse is true for 6-phosphogluconate dehydrogenase.[69] Hydrocortisone can induce glutamine synthetase activity in explant of neural retina; however, it fails to do so when dissociated retinal cells are cultured in monolayer.[70] (3) All monolayer cultures of nerve cells are derived from tumor cells and therefore may not have pertinence for normal cells. (4) Mouse neuroblastoma cells are highly aneuploid; the chromosome number per cell varies from 40 to 200,[71] while human neuroblastoma cells are closer to the normal diploid condition, with the chromosome number varying from 44 to 48.[64,65]

*Suspension Cell Culture*

Neuroblastoma cells derived from Cl300 tumor have also been grown in suspension culture. The cell line $N_2A$ was originally adapted to suspension culture conditions by Drs. Ruddle and Rosenbaum of Yale University. These cells can also be grown in monolayer culture. Cells are grown in MEM for suspension culture. The culture medium is supplemented with 5% calf serum, 5% fetal calf serum, 2 mM glutamine, 25 U/ml of penicillin G, and 10 $\mu$g/ml of streptomycin sulfate. These cells are maintained at 37° in suspension (spinner culture) in Erlenmeyer flasks by continuously stirring with a teflon-coated magnetic bar. The flasks are flushed with 5% $CO_2$ in air, and are then closed with a rubber stopper through which a small, cotton-plugged glass tube is inserted. The cell density is adjusted daily to 3 × $10^5$ cells/ml by dilution with fresh suspension culture medium.

*Advantages.* The suspension culture system has all the advantages of monolayer culture. However, it provides an additional tool in dissecting out those neuronal functions that require interaction with solid substrates from those that do not. This has been demonstrated in a recent study.[72] For example, dibutyryl cAMP induces the synthesis of a glycoprotein of molecular weight 105,000 in both suspension culture and in monolayer cells, whereas it induces the synthesis of a protein of molecular weight 78,000 only in monolayer cells.

*Disadvantages.* The suspension culture system also has the disadvantages described in the above section on monolayer cell culture. It has certain additional disadvantages. For example, the regulation of neurite formation and synapse formation cannot be studied in suspension culture. Also, many biochemical functions may not be induced if cells are grown in suspension culture.

*Culture of Whole Ganglia*

Systems that have been widely used for the *in vitro* cultivation of dorsal root and sympathetic ganglia involve either plasma clots[73] or collagen-coated culture plates.[74] These culture systems have provided much useful information with respect to neurite growth, changes in enzyme levels, formation of synapses, and effects of hormones. However, they have several limitations in regard to studying the expression of biochemical functions. For example, ganglia embedded in plasma clots cannot be harvested free, to any extent, from clot components. It is therefore difficult to measure absolute changes in the levels of specific ganglionic proteins, due to the presence of large amounts of contaminating clot proteins. In addition, drugs or hormones cannot be efficiently added and removed once the clot has formed. The harvesting of ganglia from collagen-coated plates usually requires either treatment with protease or scraping ganglia from the plates. Both of these processes may severely damage the ganglia. Mizel et al.[75] have developed a new culture technique for chick embryo dorsal root ganglion. In brief, the dorsal root ganglia are dissected from 8- and 12-day-old chick embryos, and the sympathetic ganglia are removed from 14-day-old chick embryos. Ganglia are dissected aseptically and transferred to F-12 medium, lacking serum or NGF. The sympathetic chains are separated into individual ganglia at the points of construction along the chains. The culture surface of 60-mm Falcon tissue culture dishes is overlaid with 3–5 ml of fetal calf serum for approximately 1 min. The excess serum is removed, and ganglia are transferred to the plates with a Pasteur pipette wetted with fetal calf serum to prevent ganglia from attaching to the pipette. Excess fluid is aspirated from the dishes, and each ganglion is covered separately with a very thin film of fetal calf serum. It is very important that this covering of fetal calf serum not be excessive, since the ganglia will float off the surface and remain unattached. Ganglia are then incubated at 25° for 1 hr, prior to the addition of 5 ml of medium (F-12 plus 2 $\mu$g/ml cytosine arabinonucleoside ± NGF). Dishes are incubated at 37° in a humidified atmosphere containing 5% $CO_2$. Using this method, 100 ganglia can be cultured on a single 60-mm dish.

Ganglia can also be cultured on dishes coated with a solution of 5% ovalbumin (Grade V, Sigma Chemical Corporation) in F-12 medium as described above for fetal calf serum-coated plates. However, ganglia are incubated for only 15 min with ovalbumin prior to the addition of F-12 medium containing 0.5% ovalbumin plus 2 $\mu$g/ml cytosine arabinonucleoside ± NGF.

Cytosine arabinonucleoside is used in order to prevent nonneural cell division. No inhibition of neurite outgrowth was observed at drug concentrations below 10 $\mu$g/ml.

*Advantage.* This culture system has the following advantages: (1) a large number of ganglia, 200 or more, can be cultivated and harvested with a minimum of time and materials; (2) ganglia can easily be removed intact from plates, free of exogenous materials, after a brief treatment with 0.05% EDTA in calcium- and magnesium-free balanced salt solution; (3) rapid addition of various agents to and from the culture medium, as well as washing ganglia free of the culture medium prior to harvesting, can be easily achieved; (4) since ganglia will attach and extend neurites in ovalbumin-containing medium lacking serum, the biochemical and morphological events seen after treatment with various drugs and hormones can be studied in a well-defined chemical environment.

*Disadvantage.* This technique has all the disadvantages described in the Explant Culture section.


## CONCLUSION

At least six techniques for growing nerve cells in culture are available with which to study the regulation of growth rate, differentiation, and malignancy. Each technique has its own advantages and disadvantages. Therefore, it is essential to use more than one of these techniques before making any general conclusions with respect to any one aspect of the regulation of either normal or malignant neuronal features. What molecule or molecules trigger neural differentiation? This important question in neural development remains to be answered, in spite of already extensive work. The various techniques described in this section employ cells that are relatively highly differentiated. It would be advantageous to have a monolayer culture of dividing nerve cells as well as of undetermined presumptive epidermis. The latter would provide an opportunity to study the molecular mechanisms involved in inducing neural differentiation.

It should be pointed out that all experimental model systems discussed in this section are completely removed from *in vivo* environments. The effects of adjacent tissue and circulating agents such as hormones, proteins, and ions are completely absent under culture conditions. The regulation of any given differentiated function may be controlled by more than one agent. Therefore, the identification of molecules that regulate the expression of differentiated functions in culture must be supplemented with studies on the interaction of other agents, such as hormones and ions, with these molecules in the regulation of differentiated functions in nerve cells. The effect of interaction of more than one agent in regulating any particular differentiated function may be extremely useful in our understanding of *in vivo* mechanisms of the regulation of differentiation. None of the techniques described above is suitable for the study of molecular mechanisms of learning, memory storage, and behavior patterns. For these studies, the whole organism may be the only model system that can provide information concerning these latter problems.

## REFERENCES

1. Jacobson, J., Differentiation and growth of nerve cells, in: *Differentiation and Growth of Cells in Verbebrate Tissue* (G. Goldspink, ed.), pp. 53–67, Halsted Press, New York, 1974.
2. Sato, G. (ed.), *Tissue Culture of the Nervous System*, 288 pp., Plenum Press, New York, 1973.
3. Harris, R., Allin, P., and Viza, D. (eds.), *Cell Differentiation*, 350 pp., Munksgaard, Copenhagen, 1972.
4. Pease, D. C. (ed.), *Cellular Aspects of Neural Growth and Differentiation*, 498 pp., University of California Press, Los Angeles, 1971.
5. Himwich, W. A. (ed.), *Developmental Neurobiology*, 743 pp., Charles C. Thomas, Springfield, Ill., 1970.
6. Hughes, A. F. W., *Aspects of Neural Ontogeny*, 237 pp., Elek Books/Logos Press, London, 1968.
7. Bernard, E. A. *et al.*, *Growth and Regeneration in the Central Nervous System*. Report of the Ad Hoc Subcommittee of the Advisory Council of the National Institute of Neurological and Communicative Disorders and Stroke of the National Institutes of Health, *Exp. Neurol.* **48**:1–28, 1975.
8. Balinsky, B. I., *In Introduction to Embryology*, W. B. Saunders Co., Philadelphia, 1960.
9. Bodemer, C. W., *Modern Embryology*, Holt, Rinehart and Winston, New York, 1968.
10. Saxén, L., and Toivonen, S., *Primary Embryonic Induction*, Elek Books/Logos Press, London, 1962.
11. Brachet, J., *Introduction to Molecular Embryology*, Springer-Verlag, New York, 1974.
12. Harrison, R. G., *Organization and Development of the Embryo*, Yale University Press, New Haven, 1969.
13. Niu, M. C., and Twitty, V. C., The differentiation of gastrula ectoderm in medium conditioned by axial mesoderm, *Proc. Natl. Acad. Sci. U.S.A.* **39**:985–989, 1953.
14. Barth, L. G., and Barth, L. J., The sodium dependence of embryonic induction, *Dev. Biol.* **20**:236–262, 1969.
15. Harrison, R. G., Observation on the living developing nerve fibre, *Proc. Soc. Exp. Biol. Med.* **4**:140–146, 1907.
16. Angeletti, P. U., and Levi-Montalcini, R., Two control mechanisms of growth and differentiation of the sympathetic nervous system, in: *Cellular Aspects of Neural Growth and Differentiation* (D. C. Pease, ed.), pp. 253–268, University of California Press, Los Angeles, 1971.
17. Murray, M. R., Nervous tissue *in vitro*, in: *Cells and Tissue in Culture*, Vol. 2 (E. N. Willmer, ed.), pp. 373–455, Academic Press, New York, 1965.
18. Bornstein, M. B., and Model, P. G., Development of synapse and myelin in cultures of dissociated embryonic mouse spinal cord, medulla and cerebrum, *Brain Res.* **37**:287–293, 1972.
19. Crain, S. M., Peterson, E. R., and Bornstein, M. B., Formation of functional interneuronal connections between explants of various mammilian central nervous tis-

sues during development *in vitro*, in: *Growth of the Central Nervous System* (G. E. W. Wolstenholme, ed.), pp. 13–40, Churchill, London, 1968.

20. Pomerat, C. M., Hendelman, W. J., Raiborn, C. W., Jr., and Massey, J. F., Dynamic activities of nervous tissue *in vitro*, in: *The Neuron* (H. Hyden, ed.), pp. 119–178, Elsevier, New York, 1967.

21. Weiss, P. *In vitro* experiments on the factors determining the course of the outgrowing nerve fiber, *J. Exp. Zool.* **68:**393–448, 1934.

22. Murray, M. R., and Stout, A. P., Distinctive characteristics of the sympatheticoblastoma cultivated *in vitro*. A method for prompt diagnosis, *Am. J. Anat.* **23:**120–121, 1967.

23. Bornstein, M. B., and Murray, M. R. Serial observations on patterns of growth, myelin formation, maintenance and degeneration in cultures of newborn rat and kitten cerebellum, *J. Biophys. Biochem. Cytol.* **4:**499–504, 1958.

24. Crain, S. M., and Bornstein, M. B., Early onset in inhibitory functions during synaptogenesis in fetal mouse brain cultures, *Brain Res.* **68:**351–357, 1974.

25. Bunge, R. P., Rees, R., Wood, P., Burton, H., and Ko, C. P., Anatomical and physiological observations on synapses formed on isolated autonomic neurons in tissue culture, *Brain Res.* **66:**401–412, 1974.

26. Scott, B. S., Effect of potassium on neuron survival in cultures of dissociated human nervous tissue, *Exp. Neurol.* **30:**297–308, 1971.

27. Phillipson, O. T., and Sandler, M., The influence of nerve growth factor, potassium depolarization and dibutyryl (cyclic) adenosine 3′,5′-monophosphate on explant culture of chick embryo sympathetic ganglia, *Brain Res.* **90:**273–281, 1975.

28. Basrur, P. K., Basrur, V. R., and Gilman, J. P. W., A simple method for short term cultures from small biopsies, *Exp. Cell Res.* **30:**229–235, 1963.

29. Levi-Montalcini, R., and Angeletti, P. U., Essential role of the nerve growth factor in the survival and maintenance of dissociated sensory and sympathetic embryonic nerve cells *in vitro*, *Dev. Biol.* **7:**653–659, 1963.

30. Moscona, A. A., Recombination of dissociated cells and the development of cell aggregates, in: *Cell and Tissue Culture* (E. N. Willmer, ed.), pp. 489–529, Academic Press, New York, 1965.

31. Moscona, A. A., Surface specification of embryonic cells: lectin receptors, cell recognition and specific cell ligands, in: *The Cell Surface in Development* (A. A. Moscona, ed.), pp. 67–99, Wiley, New York, 1974.

32. Delong, G. R., Histogenesis of fetal mouse isocortex and hippocampus in reaggregating cell cultures, *Dev. Biol.* **22:**563–583, 1970.

33. Sidman, R. L., and Wessells, N. K., Control of direction of growth during the elongation of neurites, *Exp. Neurol.* **48:**(3)Part 2:237–251, 1975.

34. Seeds, N. W., Differentiation of aggregating brain cell cultures, in: *Tissue Culture of the Nervous System* (G. Aato, ed.), pp. 35–53, Plenum Press, New York, 1973.

35. Seeds, N. W., Gilman, A. G., Amano, T., and Nirenberg, M. W., Regulation of axon formation by clonal lines of a neural tumor, *Proc. Natl. Acad. Sci. U.S.A.* **66:**160–167, 1970.

36. Trinkaus, J. P., and Groves, P. W., Differentiation in culture of mixed aggregates of dissociated tissue cells, *Proc. Natl. Acad. Sci. U.S.A.* **41:**787–795, 1955.

37. Moscona, A. A., Development of heterotypic combinations of dissociated embryonic chick cells, *Proc. Soc. Exp. Biol. Med.* **92:**410–416, 1956.

38. Garber, B. B., and Moscona, A. A., Reconstruction of brain tissue from cell suspension. I. Aggregation patterns of cells dissociated from different regions of the developing brain, *Dev. Biol.* **27:**217–234, 1972.

39. Morris, J. E., and Moscona, A. A., The induction of glutamine synthetase in cell aggregates of embryonic neural retina: Correlations with differentiation and multicellular organization, *Dev. Biol.* **25**:420-444, 1971.

40. Varon, S., and Raiborn, C. W., Jr., Dissociation, fractionation and culture of embryonic brain cells, *Brain Res.* **12**:180-199, 1969.

41. Varon, S., and Raiborn, C., Dissociation, fractionation and culture of chick embryo sympathetic ganglion cells, *J. Neurocytol.* **1**:211-221, 1972.

42. Lowry, O., The chemical study of single neurons, *Harvey Lect.* **58**:1-19, 1963.

43. Booher, J., and Sensenbrenner, M., Growth and cultivation of dissociated neurons and glial cells from embryonic chick, rat and human brain in flask cultures, *Neurobiology* **2**:97-105, 1972.

44. Hydén, H., Cytophysiological aspects of the nucleic acids and protein of nervous tissue, in: *Neurochemistry* (K. A. C. Elliott, I. H. Page, and J. H. Quastel, eds.), pp. 331-375, Charles C Thomas, Springfield, Ill., 1962.

45. Roots, B. I., and Johnston, P. V., Neuron of ox brain nuclei: Their isolation and appearance by light and electron microscopy, *J. Ultrastruct. Res.* **10**:350-361, 1964.

46. Cohen, A. I., Nicol, E. C., and Richter, W., Nerve growth factor requirement for development of dissociated embryonic sensory and sympathetic ganglia in culture, *Proc. Soc. Exp. Biol. Med. (N.Y.)* **116**:784-789, 1964.

47. Nakai, J., Dissociated dorsal root ganglia in tissue culture, *Am. J. Anat.* **99**:81-99, 1956.

48. Varon, S. S., Raiborn, Jr., C. W., Seto, T., and Pomerat, C. M., A cell line from trypsinized adult rabbit brain tissue, *Z. Zellforsch Mikrosk. Anat.* **59**:35-46, 1963.

49. Rinaldini, L. M. J., The isolation of living cells from animal tissue, *Int. Rev. Cytol.* **7**:587-647, 1958.

50. Korey, S. R., Some characteristics of a neuroglial fraction, in: *Metabolism of Nervous System* (D. W. Richter, ed.), pp. 87-90, Pergamon Press, New York, 1957.

51. LeBaron, F. N., Determination of protein bound phosphoinositides in glial cell concentrates of brain white matter, in: *Variation in Chemical Composition of the Nervous System as Determined by Developmental and Genetic Factors* (G. B. Ansell, ed.), p. 65, Pergamon Press, New York, 1966.

52. Chu, L. W., A cytological study of anterior horn cells isolated from human spinal cord, *J. Comp. Neurol.* **100**:381-399, 1954.

53. Rose, S. P. R., Preparation of enriched fractions from cerebral cortex containing isolated, metabolically active neuronal and glial cells, *Biochem. J.* **102**:33-43, 1967.

54. Satake, M., and Abe, S., Preparation and characterization of nerve cell perikaryon from rat cerebral cortex, *J. Biochem. (Tokyo)* **59**:72-75, 1966.

55. diZerega, G., Johnson, L., Morrow, J., and Kasten, F. H., Isolation of viable neurons from embryonic spinal ganglia by centrifugation through albumin gradients, *Exp. Cell. Res.* **63**:189-192, 1970.

56. Prasad, K. N., Differentiation of neuroblastoma cells in culture, *Biol. Rev.* **50**:129-165, 1975.

57. Dunham, L. T., and Stewart, H. L., A survey of transplantable and transmissible animal tumors, *J. Natl. Cancer Inst.* **13**:1299-1377, 1953.

58. Augusti-Tocco, G., and Sato, G., Establishment of functional clonal lines of neurons from mouse neuroblastoma, *Proc. Natl. Acad. Sci. U.S.A.* **64**:311-315, 1969.

59. Schubert, D., Humphreys, S., Baroni, C., and Cohn, M., *In vitro* differentiation of a mouse neuroblastoma, *Proc. Natl. Acad. Sci. U.S.A.* **64**:316-323, 1969.

60. Buonassisi, V., Sato, G., and Cohen, A. I., Hormone-producing cultures of adrenal and pituitary tumor origin, *Proc. Natl. Acad. Sci. U.S.A.* **48**:1184-1190, 1962.

61. Puck. T. T., Marcus, P. I., and Cieciura, S. J., Clonal growth of mammalian cells *in vitro*. Growth characteristics of colonies from single HeLa cells with and without a "Feeder" layer, *J. Exp. Med.* **103**:273–284, 1956.
62. Prasad, K. N., and Vernadakis, A., Morphological and biochemical study in X-ray and dibutyryl cyclic AMP-induced differentiated neuroblastoma cells, *Exp. Cell Res.* **70**:27–32, 1972.
63. Prasad, K. N., Mandal, B., Waymire, J. C., Lees, G. J., Vernadakis, A., and Weiner, N., Basal level of neurotransmitter synthesizing enzymes and effect of cyclic AMP agents on the morphological differentiation of isolated neuroblastoma clones, *Nature (London) New Biol.* **241**:117–119, 1973.
64. Tumilowicz, J. J., Nichols, W. W., Cholon, J. J., and Greene, A. E., Definition of a continuous human cell line derived from neuroblastoma, *Cancer Res.* **30**:2110–2118, 1970.
65. Biedler, J. L., Helson, L., and Spengler, B. A., Morphology and growth, tumorigenecity and cytogenetics of human neuroblastoma cells in continuous culture, *Cancer Res.* **33**:2643–2652, 1973.
66. Goldstein, M. N., Burdman, J. A., and Journey, L. J., Long-term tissue culture of neuroblastomas. Morphologic evidence for differentiation and maturation, *J. Natl. Cancer Inst.* **32**:165–199, 1964.
67. Schubert, D., Heinemann, S., Carlisle, W., Tarikas, H., Kimes, B., Patrick, J., Steinbach, J. H., Culp, W., and Brandt, B. L., Clonal cell lines from the rat central nervous system, *Nature (London)* **249**:224–227, 1974.
68. Nelson, P., Christian, C., and Nirenberg, M., Synapse formation between clonal neuroblastoma × glioma hybrid cells and striated muscle cells, *Proc. Natl. Acad. Sci. U.S.A.* **73**:123–127, 1976.
69. Prasad, N., Prasad, R., and Prasad, K. N., Electrophoretic patterns of glucose metabolizing enzymes and acid phosphatase in mouse and human neuroblastoma cells, *Exp. Cell Res.* **104**:273–277, 1977.
70. Morris, J. E., and Moscona, A. A., Induction of glutamine synthetase in embryonic retina; its dependence on cell interactions, *Science* **167**:1736–1738, 1970.
71. Amano, T., Richelson, E., and Nirenberg, M., Neurotransmitter synthesis by neuroblastoma clones, *Proc. Nat. Acad. Sci. U.S.A.* **69**:258–263, 1972.
72. Truding, R., Shelanski, M. L., Daniels, M. P., and Morrel, P., Comparison of surface membranes isolated from cultured murine neuroblastoma cells in the differentiated or undifferentiated state, *J. Biol. Chem.* **249**:3973–3982, 1974.
73. Levi-Montalcini, R., Meyer, H., and Hamburger, V., *In vitro* experiments in the effects of mouse sarcomas 180 and 37 on the spinal and sympathetic ganglia of the chick embryo, *Cancer Res.* **14**:49–57, 1954.
74. Roisen, F. J., Murphy, R. A., and Braden, W. G., Neurite development *in vitro*. I. The effects of adenosine 3′,5′-cyclic monophosphate cyclic AMP, *J. Neurobiol.* **3**:347–368, 1972.
75. Mizel, S. B., and Bamburg, J. R., Studies on the action of nerve growth factor. I. Characterization of a simplified *in vitro* culture system for dorsal root and sympathetic ganglia, *Dev. Biol.* **49**:11–19, 1976.

# 2

# NEURAL INDUCTION

## INTRODUCTION

The phenomenon of induction is one of the most important concepts in embryonic development. Embryonic induction may be defined as the process by which one group of cells and/or cell products causes a second group of cells to differentiate into cells which differ from parent cells. For example, the roof of archenteron causes the overlying ectoderm to differentiate into neural tissue. Modification of genetic and structural characteristics occurs at each induction. Many successive inductions are involved in the development of embryonic structures. The embryo is a dynamic system in which the topographic relations of cells and cell groups are in constant flux. The spatial and temporal order of their relations are critical to the normal development and to the nature of inductive interactions at any given point and time. For example, the original archenteron roof does not remain in contact with the same ectoderm throughout gastrulation; the above structures progressively change their character. The reactive ability of the presumptive ectoderm of the gastrula diminishes with time, and this reactivity is influenced by exposure to previous inductive influences. An initially reacting tissue may even acquire inducing capacity. For example, the neural plate, once induced, is itself capable of inducing early gastrular ectoderm to form neural plate. The optic vesicles also perform an inductive role following their own induction. The phenomenon of successive induction emphasizes the multiplicity and diverse quality of embryonic inductive reactions. Many inducing systems operate during the course of development; one of these, referred to as Spemann's organizer, is responsible for the formation of the nervous system.

### Spemann's Theory of the Organizer

The term "organizer"[1] initially referred to a limited area of the dorsal lip of the blastopore (two-celled newt egg stage). If the blastopore halves were separated along the median plane, both halves developed into complete embryos. If the two halves were separated along a plane at a right angle to the median plane, only the dorsal half formed a complete embryo. Spemann also demonstrated the dependence of lens formation upon the presence of an optic vesicle.[1] For example, when the prospective eye rudiment was removed from grass frog embryos, a lens did not develop; conversely, an optic vesicle transplanted under the belly ectoderm caused formation of lens instead of belly skin. These are important findings, suggesting that optic vesicles contain specific chemical agents capable of influencing the differentiative outcome of presumptive ectoderm or of the groups of cells necessary for inducing such an event. Spemann also showed that the exchange of small pieces of presumptive neural plate and skin ectoderm between early salamander gastrulae led each tissue to develop according to its new location, whereas a piece of the dorsal lip of the blastopore developed independently of its new surroundings, into an embryolike body. In neural induction, two factors are required: the inducing *stimulus* coming from the organizer, and the *competence* of the reacting ectoblast. If the ectoblast is taken from an aged gastrula, it will have lost the capacity to react to the inducing stimulus coming from the organizer. It will no longer be competent to react and will form only epidermis.

### NEURAL INDUCTION

The elegant studies of such early embryologists as Needham, Waddinton, Finkelstein, Brachet, Fischer, Wehmeier, Holtfreter, Barth, Lehman, Shen, Abercrombie, Woerdeman, Blansky, Shapiro, Yamada, and Tiedeman have provided a great deal of information concerning artificial experimental conditions under which the induction of neural tissue will occur. However, the mechanisms by which induction of neural tissue in the whole organism occur remain unknown. The following possibilities have been suggested: (1) surface interaction; (2) chemical mediation; and (3) cellular death. A great deal of evidence suggests that a diffusible substance may act as an effective inductive stimulus. When a piece of dorsal blastopore lip was separated from a piece of presumptive early gastrula ectoderm by a Millipore filter (20 $\mu$m thick with an average pore size of 8 $\mu$m), the treated ectoderm formed a vesicle with definite brain and eye. Under these conditions large molecules could traverse the filter, but the inducing tissues were not in physical contact.[2] An additional experiment supports the concept that the inducing substances are diffusible materials elaborated by inducer cells. A small portion of the dorsal lip of the blastopore of an amphibian gastrula was removed and placed in a small volume of suitable medium. After 7–10 days, these cells were removed and discarded. The medium in which the cells had been maintained is referred to as "conditioned medium."[2,3] When a small fragment (10–20 cells) of ectoderm was removed from an early gastrula and placed in the conditioned medium, the ectoderm developed into nerve and pigment cells.

It has also been shown[2] that the inductor tissue may contain more than one inductor substance. For example, the implantation of HeLa cells into early salamander embryo tail causes formation of spinal cord and mesodermal structures. Thus, HeLa cells contain inductors for both mesoderm and nerve cells. In spite of the demonstration of inductive phenomena in neural tissue, the nature of the inducing substances remains unknown.

## Nature of Inducing Substance

The fact that the dead dorsal lip retained the inducing capacity[3] shows that the inducing capacity of the organizer is due to a chemical substance. The methods of inactivating dorsal lip generally involved heating, drying, or fixing the tissue in alcohol. In addition to dead dorsal lip, a piece of dead ectoderm (killed by boiling) also maintains the capacity to induce nervous tissue.[3] The viable ectoderm is not an inductor. This finding was initially interpreted to mean that the ventral fragments contained an inhibiting substance destroyed upon killing the tissue.[3] This hypothesis was discarded because the dead organizer stimulated induction when enclosed in a sheet of isolated ventral epidermis. The dead ectoderm or endoderm was equally effective in inducing nervous tissue. Thus, the factor responsible for induction may be a chemical substance diffusing from the dead tissue.

The inducing substance was resistant to boiling (1 hr) and to either, xylol, and 20% HCl (for 24 hr). However, heating at 100° for 1.5 hr or at 185° for 1 hr completely inactivated the inducing substance.[3] Nervous systems of considerable size were formed when different types of organs from different species, either dead or alive, were implanted into the gastrula. The inducing capacity of various organs is summarized below:

Very powerful inductors
    1. Coagulated chick embryo extract
    2. Liver, kdiney, adrenal, heart, brain of mouse
    3. Thyroid, kidney, liver, brain, tooth of man
    4. Bottom layer of centrifuged brei of calf liver
    5. Liver of lizard, frog, triton
    6. Brain and retina of salamander
    7. Heart, ovarian egg, muscle, liver of fish

Powerful inductors
    1. Lens of mouse
    2. Centrifuged calf-liver brei (middle and upper layers)
    3. Thyroid, kidney, liver, testicle, adrenal, fat of birds
    4. Kidney and testis of lizard
    5. Heart and limb bud of triton
    6. Frog muscle
    7. Dragonfly ganglion substance
    8. Lymph from *Sphinx* larva
    9. *Limnaea* liver
   10. *Dalphnia* extract

Weak inductors
    1. Blood and fatty tissue of mouse
    2. Liver and heart of salamander
    3. Retina of triton larva
    4. Fatty tissue from dragonflies
    5. Imaginal disk from caterpillar of *Vanessa*
    6. Foot muscle of *Planorbis* and *Limnaea*
    7. Subcutaneous muscle of *Enchytrids*

Inactive inductors
    1. Ectoderm and endoderm of living gastrula
    2. Tadpole gills
    3. Starch, agar, egg albumin, pork fat, animal charcoal

Later studies[3] concluded that heterogenetic inductors (dead tissue) never induce normal organ structure with respect to histological appearance and typical anatomical relationships. The implanted dead tissue does not induce perfect morphogenesis; it merely stimulates new potentialities in contacted regions. It was pointed out that the living organizer stimulates the normal morphogenesis of nervous system in a manner exhibiting regional characteristics, forming a cerebral vesicle anteriorly, followed by a neural tube, the dimensions of which gradually decrease posteriorly. The picture is quite different after implantation of a dead organizer or of certain other organs. These heterogenetic inductors induce nervous tissues which are complex and abnormal. For example, a cerebral vesicle, a notochord, and eyes may be found mixed together. Two different terminologies were therefore proposed.[3] The term *evocation* was used when the induction consisted of an atypical mass, and the term *individuation* was used when the host formed a typical organ which differentiated along the normal embryonic axis. Heterogenetic inductors are thus evocators rather than organizers. The organizer term should be reserved exclusively for the living dorsal lip of the blastopore, since it alone is capable of inducing nervous system tissue perfect regional organization, i.e., individuation.

When an evocator is implanted, it undergoes chemical changes during its contact with the host. If living tissue is implanted, the evocator soon becomes autolyzed; and if a dead organ is introduced, the evocator may be rapidly destroyed by the hydrolytic enzymes of the host. Therefore, the active substance is either present initially in the evocator, or the active substances are formed secondarily by autolysis or enzymatic digestion. The fragments of dead organs lost all activity when covered with a layer of agar or gelatin that protected them from the action of the neighboring host cells.[3] The autolysates of various organs, notably the hypophysis, are evocators after being incorporated into agar.[3] These observations lead one to suspect that in such cases the evocating agent is a soluble product of enzymatic hydrolysis. Indeed, it was demonstrated that the dead tissues were attacked by host cells. The fragments of the organizer or of various organs fixed by either boiling or alcohol treatment did not give the cytochemical reaction for thymo- and ribonucleic acids a few days after their implantation into the gastrula. There was little doubt that the nuclease of the host had hydrolyzed the nucleic acid of the graft. Substances other than nucleic acids might also have been hydrolyzed.

Extracts of ground-up neurulae proved to be excellent evocators.[3] When a brei of embryos was centrifuged, three layers were seen: a fatty top layer, a liquid containing proteins and fine granules, and a bottom layer of yolk. The implantation of the fatty layer and the yolk layer gave positive induction. However, the middle layer, after coagulation of the proteins by heat, gave induction in only 16% of the cases. Ether extract or petrol ether extract of the embryo or adult tissue extract gave an 8% rate of induction. The lipid remained active after saponification, suggesting that the evocator substance was soluble in ether or petroleum ether, and was possibly a *sterol*. However, a few pure sterols such as cholesterol, oestrol, calciferol, and phytosterol do not produce induction of nervous tissue. The quality of induction obtained with aqueous extracts, petroleum ether extract, and the unsaponifiable fraction of neurulae is inferior. The induction is characterized by the complete absence of mesodermal inductions (notochord or somites). Whenever living and dead tissues or the normal organizers are used as inductors, notochord and somites are frequently found together with the nervous tissue.

The possibility that the inducing substance is a glycogen was discarded after further experimentation showed that chick embryo extract, which contains no glycogen, is an effective inducing agent.

An ether extract of impure glycogen gives good neural induction in a number of cases. In addition, when the sterols from glycogen were precipitated by digitonin, a preparation that stimulated induction was obtained. Thus, the lipid nature of the evocator remains a possibility. It was later shown[3] that the synthetic polycyclic hydrocarbons were also evocators. The secondary nervous systems, however, were generally small and atypical. Data supported the sterol theory, since the hydrocarbons used were chemically similar to the sterols.

It was also suggested that the evocator was not a sterol but rather a phosphatide, cephalin. However, lipid free of cephalin maintains its inducing capacity; therefore the possibility remains that the inducing substance of an evocator is a sterol. The inductive capacity of dead organs was not impaired when they were freed from lipid by treatment with boiling fat solvent. The ether extract of the embryo containing evocators loses its inducing activity if the extract is neutralized with bicarbonate. These results led to the belief that the inducing substance is an acid, soluble in ether and in water. However, implants of neutral fats, phosphatides, and the unsaponifiable fraction produced no induction. On the other hand, fatty acids isolated from triglycerides and synthetic oleic acid were active. Later studies[3] showed that fatty acids are not the only agents responsible for evocation by dead organs, since the residue after removal of fats remains extremely active.

Others have suggested[3] that the inducing substances are nucleoproteins. The nucleoproteins from thymus, pancreas, and liver are excellent inductors even after prolonged treatment to remove fats and after purification by several precipitations and by dialysis. The nucleic acids of thymus and of pancreas are also excellent inductors even after elimination of lipids. It was also observed that nucleic acids and nucleoproteins induce neural structures only when implanted in their natural state. If they are embedded in agar, their inducing activity disappears. It is thus possible that nucleic acids are first hydrolyzed by the host enzyme into simpler and more soluble derivatives, probably mononucleotides. Indeed, adenylic acid from muscle has been found to be a good inductor. Other mononucleotides failed to show any inducing activity. These experiments led to the conclusion that only acid substances are able to induce the formation of a nervous system in the ectoderm, and it makes no difference whether the acid is fat- or water-soluble. Induction by dead tissue may then be due to stimulation by an acid; whether or not this mechanism plays a role in normal induction remains to be determined.

The nucleoproteins that were slightly acid and completely lipid-free were inductive in about 73% of cases.[3] When thymonucleic acid was separated from histone by means of alkali, the nucleic acid induced nervous tissue in 51% of cases. However, histone not completely free of nucleic-acid-induced nervous tissue in only 15% of cases. A thorough purification of the thymonucleic acid reduced its activity; but an alkaline hydrolysis, which increases the amount of free nucleotides, restored the entire inducing activity. Although thymonucleic acid did not induce nervous tissue when embedded in agar, adenylic acid from muscle, under similar experimental conditions, retains its inducing activity. When nucleotides prepared from any organ were subjected to increasing purification, a progressive decrease in inducing activity occurred. This may be due to the gradual elimination of the proteins, resulting in an excessive solubility of the nucleotides. Guanylic acid, inosinic acid, inosine, and zymonucleic acid showed no inducing activity. Thus, there were two ideas regarding the nature of inducing substances in an evocator: that the inducing substances were either acid derivatives or sterols.

It was shown[3] that the host may play a role in the induction mechanism. According to this concept, chemical agents (evocators) merely act as stimulators or activators, unable to create new potentialities in the host. These agents simply influence the various regions of the host by favoring a reaction. Thus, the embryo can be divided into a series of fields. For example, when the optic vesicle is implanted on the ectoderm, the ectoderm forms a lens. However, the efficiency of ectoderm to form a lens under the above experimental condition is not the same at all points; it decreases gradually with the distance from the lens primordium. A similar situation exists in other regions. The presumptive neural plate reacts more readily with the organizer to form a nervous system than does the presumptive epidermis. Heterogenetic inductors give a better neural reaction if during gastrulation it comes to lie near the nervous system of the host. The induced neural tube will be better-developed in the anterior part than in the posterior part, which is similar to that observed with the normal system. However, it should be pointed out that one obtains well-formed neural tubes at regions far from the host nervous system. Nevertheless, the induction reaction may in part depend upon the host region.

Glutathione, cysteine, and tryptophan mixed with agar, dinitrophenol, thyroxin, vitamins (A, B, $B_2$, C, $D_2$), adrenalin, and heteroauxin were inactive in inducing nervous tissue.[3]

It was pointed out[3] that inducing substances may be protein. It was shown that the best evocator is the precipitate obtained when gastrulae were extracted with NaOH at low temperature and brought to pH 5 with acetic acid. This preparation, chiefly of protein nature, gives normally appearing nervous systems in a high percentage of cases. If the protein is denatured, it loses its ability as an inductor and becomes a simple evocator. The activity of protein isolated from neuralae by different methods was compared for their capacity to induce nervous tissue.[3] The methods of isolating protein involved extraction with concentrated NaCl and precipitation by dilution, extraction by dilute NaOH and precipitation with acetic acid, and shaking with chloroform. Only the NaOH extract showed good activity.

It was suggested[3] that -SH groups may participate in the induction of nervous tissue. This mechanism may explain why noninductor regions of the gastrula become evocators when they are killed. During denaturation of the proteins there is an increase in the number of -SH groups. Cytolyzing agents also produce a partial denaturation, with the liberation of free -SH groups. Thus, one can comprehend the effectiveness of the sterols and acids of this mechanism. In addition, there appears to be a parallel between the amount of -SH groups in various organs and their inducing power. For example, in amphibian ovary, young oocytes are better inductors than old ones,[3] and their -SH content is much higher. In large oocytes, the germinal vesicle rich in -SH groups is distinctly more effective than the cytoplasm. However, some observations fail to support the concept of -SH groups having a role in the induction of nervous system. For example, egg albumin is not an evocator, although it has large amounts of -SH groups. The administration of -SH-containing amino acids such as cysteine and glutathione did not induce nervous tissue. The blocking of -SH groups is not adequate to destroy the evocator action of the various protein fractions.[3] Thus, the -SH groups do not seem to be important factors in producing nervous tissue by heterogenetic inductors. However, this role in the normal induction cannot be ruled out. It has been suggested[3] that -SH groups are necessary mainly for the reaction of the presumptive neural plate to the inductor. This idea is supported by the fact that embryos in which the -SH groups are blocked by iodoacetamide or choloropicrin form a notochord and normal somites, but have only rudimentary nervous systems, often incompletely closed. It thus seems that the mechanisms insuring the thickening of the neural plate, its folding into a trough, and its closure require the presence of -SH groups.

The possible participation of ribonucleoprotein in the induction of neural tissue by heterogentic inductors is strengthened by the suggested role of -SH groups. This also explains the failure of egg albumin, cysteine, and glutathione in inducing neural tissue, since these substances do not contain nucleic acid. It has been shown[3] that an operation results in evocation only when the nucleic acids present in the implant are attacked by the host. This was further substantiated by the observation that fragments of dead organizer, when treated with pure ribonuclease, fail to induce any neural tissue.

The question of the role of RNA in evocative activity was further investigated. It was found that the frequency of evocation is, in large part, proportional to the RNA content of the implants. RNA itself is not necessarily the inductor, but RNA hydrolyzed into more diffusible low-molecular-weight derivatives by the ribonuclease of the host. One of the nucleotides may therefore be the active factor. The evocative ability of heterogenetic inductors, in particular that of dead organs, may be due to their nucleic acids, which undergo degradation in the host.[3]

It has also been reported[4] that a simple shift in the pH of the salt solution in which the pieces of ectoderm were cultured is enough to produce numerous cases of spontaneous transformation of the ectoblast into nerve cells. In addition, lithium chloride, sodium chloride, $CO_2$, ammonia, crushed glass, and Teflon were also good inducers of nervous-system formation. These data suggested that the organizer might act by producing these simple end products or its metabolites. They would act by a "release" mechanism and liberate an unknown "neurogenic" substance from an inactive complex present in the cells of the competent ectoblast. In 1950, Yamada and Teideman reinvestigated the problem of induction of neural tissue. They selected amphibian species in which spontaneous neuralization of explanted ectoblast is difficult to obtain, and in which the introduction of $CO_2$ into the culture medium is insufficient to transform ectoblast into nerve cells. They soon isolated proteins with different inducing specificities. A soluble protein, which can be extracted and purified from bone marrow and form chicken embryos, acts as a *mesodermalizing* factor, transforming gastrula ectoblast into mesodermic tissue, such as chorda or muscle. The ectoblast that normally would have formed epidermis (or nerve cells, if in the presence of an inducer) becomes a tail, partially or completely devoid of a nervous system. Another protein, associated with the ribosomes of chick embryos, is a *neuralizing* factor. At low concentrations, it very effectively transforms the ectoblast cells into neural cells. A third protein is an inhibitor of the mesodermalizing factor and might act to control the activity of this factor.

During neural induction, the synthesis of DNA as evidenced by increased mitotic activity as well as increased levels of RNA and protein, markedly increases.[4] The addition of neuralizing and mesodermalizing factors into the medium containing fragments of ectoblasts induced an increase in nucleic acid and protein synthesis. Thus, neural induction is accompanied by a marked increase in gene expression. If an organizer is treated with actinomycin D (inhibits RNA synthesis) and placed in contact with normal ectoblast, neural induction occurs. If, on the other hand, the normal organizer is placed in contact with actinomycin-D-treated ectoblasts, no neural induction is observed. This shows that the ectoblast is unable to respond to an inducing stimulus if its RNA synthesis is blocked. However, the organizer retains its inducing activity even in the absence of RNA synthesis. The effect of puromycin, an inhibitor of protein synthesis, has not been extensively investigated. It has been shown[4] that microcephaly occurs frequently in embryos treated with puromycin. Both colchicine (which inhibits assembly of microtubules) and cytochalasin B (which destroys microfilaments) inhibit neurulation in an amphibian egg.[4]

Cell movements play an important part in early neurulation, when the neural tissue first forms a flat plate that then changes into a groove and finally into a tube. This transformation of neural plate into neural tube is completely blocked by mercaptoethanol (which suppresses the mitotic apparatus). On the other hand, ATP markedly accelerates both the formation of neural tube from neural plate and the closure of the neural tube. The exact mechanisms by which ATP and mercaptoethanol act are unknown. These agents are known to produce many metabolic effects.

When an organizer of one species is grafted into a gastrula of another species, normal head with eyes, nose, and so on, can be induced in the host, but this secondary head is eventually rejected and destroyed. Thus, the inducing capacity of an organizer does not display any species specificity, but the response of the host to the inductor is species-specific, i.e., gene-determined. A frog organizer grafted into a newt embryo induces newt neural tissue. The immune response of amphibian embryo is different from that of mammalian embryo. If a foreign tissue is grafted into the mammalian embryo, it is not rejected, even when the immune system is developed after birth. The reasons for this difference are not known.

The problem of differentiation of specialized cells from the same source also remains poorly understood. For example, neural crest cells give rise to melanocytes as well as nerve cells. It is known[4] that when a piece of neural crest is cultured in vitro, both types of cells are formed. The addition of phenylalanine into the culture of neural crest stimulates the differentiation of neural crest into melanocytes.[4] On the other hand, addition of an analogue of phenylalanine, fluorophenylalanine, inhibits the formation of melanocytes without interfering with that of nerve cells. The differentiation of neural crest cells can thus be switched in either of two directions. It should be pointed out that melanocytes can lose the capacity to synthesize melanin in a reversible way. For example, when melanocytes are cultured in vitro, they become colorless after a few divisions. However, when they come in contact with each other, they synthesize melanin again. Such cases of differentiation and redifferentiation cannot readily be explained by theories which assume that the DNA molecules themselves undergo chemical modification during cell differentiation. It would be important to identify factors that regulate the expression of specific genes in cells of higher organisms during induction processes.

Although many substances have been shown to induce neural tissue in the ectoblast,[3] the best evidence to date suggests that the primary inductors under physiological conditions may be proteins.[5] The role of adenosine 3',5'-cyclic monophosphate (cAMP) in the induction mechanism became a possibility when it was discovered[6-8] that an elevation of cAMP in neuroblastoma cells in culture induces as well as increases the expression of many differentiated functions characteristic of mature neurons. Indeed, it has been shown[9] that dibutyryl cAMP, 8-bromo cAMP, and cAMP with theophylline can induce undetermined cells to differentiate into derivatives of neural ectoderm and neural crest (neurons, melanophores and glial cells). Other related nucleotides such as 5'-AMP, 2',3-AMP, dibutyryl cGMP, and sodium butyrate were ineffective.[9] The results obtained varied somewhat from one species to another. Neural differentiation was observed in all experimental explants; however, on rare occasion the formation of simple neurons was observed in control explants. Generally, the following recognizable cell types were observed with increasing duration of incubation: simple motor neurons, astrocytes, melanophores, and central nervous system neurons. In addition, structures resembling nerve bundles were also observed in some cultures of *Siredon mexicanum* cells. The rate of formation of these cell types varies with the animal used, being much more rapid with *Xenopus laevis* than with *Pleurodeles waltili* or *Siredon mexicanum*. The estimated percentage of cells that developed into neural ectoderm-derived cells varied from 10 to 50.

Theophylline alone induced limited neural differentiation only in case of *P. waltili*. cAMP alone gave variable results, usually with limited neural differentiation.[9] This is expected, since theophylline has failed to increase cAMP in several types of cells,[10] although it inhibits cyclic nucleotide phosphodiesterase activity *in vitro*. cAMP is known to penetrate intact cells poorly.

It has been reported[9] that two different analogs of cAMP produced quantitatively different results. For example, in *S. mexicanum* 8-bromo cAMP caused the formation of large numbers of astrocytes with few simple motor neurons and practically no central nervous system neurons, whereas dibutyryl cAMP caused the formation of both types of neurons as well as astrocytes.

Dibutyryl cAMP induces neural differentiation in small explants of presumptive epidermis, but fails to do so in large explants or in whole embryo.[11] Dibutyryl cAMP, melanocyte-stimulating hormone (MSH), cAMP and theophylline, or theophylline alone accelerated the cytodifferentiation (melanin synthesis) of melanophore in whole embryos or in large explants. But these same compounds cannot induce melanophore in large explants of presumptive neural ectoderm unless they have been induced to form melanophore by invaginating choramesoderm.[12] It is well known that the process of neural differentiation involves determination as well as cytodifferentiation (development of morphological characteristics) and biochemical differentiation. An extensive morphogenetic movement also occurs *in vivo* during gastrulation and neuralization. This is striking in the case of melanophores, as cytodifferentiation occurs only after extensive migration of the determined melanoblasts.[13] Dibutyryl cAMP may fail to induce neural differentiation in large explants in culture because they roll into a ball and probably do not undergo proper morphogenetic movement. However, in the case of small explant culture, the explants attach to the plastic surface, and cells migrate out to eventually become a monolayer. It has also been observed[11] that neural differentiation was always more pronounced in the area surrounding the explants where cells had become isolated or in loose contact with each other. Based on these observations, the following hypotheses have been proposed[11]: (1) Dibutyryl cAMP can induce determination; (2) the induced cells are inhibited from cytodifferentiation by cell contact with adjacent cells or with adjacent inhibitory cells; and (3) cellular movements separating the determined cells from adjacent cells or from adjacent inhibitory cells allow cytodifferentiation to take place. The cellular movements are known to occur in the presumptive neural ectoderm *in vivo;* however, in small explant cultures this is achieved by the migration of cells on the surface of the culture dish. To test this hypothesis, an elegant and simple experiment was performed.[11] Large explants of presumptive epidermis were first incubated in the present of dibutyryl cAMP for 12 to 24 hr, and were then washed. The explants were then dissociated into small pieces and were cultured in the absence of dibutyryl cAMP. It was found that the cells migrated out from the explant and underwent neural differentiation. These data show that neural induction had taken place during the period of incubation of explants with dibutyryl cAMP. Once the induction had taken place, external cAMP was no longer required for the cytodifferentiation of the induced cells. It was also observed that cells from small explants not treated with dibutyryl cAMP also migrated away from each other. However, no neural differentiation occurred. Thus, cell migration *per se* is not sufficient for neural differentiation. Apparently a combination of induction by cAMP, followed by cell migration in culture, is essential for neural differentiation. This may also explain why Yamada[14] has been able to achieve much more efficient induction of neural differentiation by keeping the large explants flat with a nylon mesh, creating a different topology for cell migration as compared to a ball of cells.

It has been admitted[9] that these results do not in themselves prove that cAMP is the normal agent mediating the effect of primary inductors. Before cAMP involvement in the induction mechanism is considered, at least one stimulator of adenylate cyclase must induce neural induction. The intracellular level of cAMP in neural tissue after normal induction must increase. It is interesting to note[15] that during mesoderm segmentation the cAMP level increases in various regions of the chick embryo.

## Effect of Ions

It has been reported[16] that brief treatment with high concentrations of cations of monolayer cell cultures derived from presumptive epidermis of *Rana pipiens* resulted in the sequential appearance of cells characteristic of neural fold origin.

Barth and Barth[17] have proposed a working hypothesis for the mechanism of neural induction as follows. Inducting compounds, both natural and synthetic, have in common an alteration in cell membrane properties, resulting primarily in the release and redistribution of inorganic ions to new binding sites. The resulting switch in type of differentiation is considered to be based initially upon differential distribution of cytoplasmic components in the ovum. It has been demonstrated[18] that the actual induction process may be initiated by the release of bound ions. This concept need not involve changes in total ion concentrations, but rather a change in the ratio between bound and free ions within the cells of early gastrulae.

Naturally occurring cations, such as calcium chloride (2.4 mM) and magnesium sulfate (12.8 mM), induced motor and ganglion nerve patterns in cultures of presumptive epidermis. These cation concentrations are well below those which cause osmotic shock. After induction, the treated aggregates may be cultured continuously in 7.47 mM $CaCl_2$ or 20.8 mM $MgSO_4$ added to standard solution.

Inductive stimuli, such as sucrose, calcium, lithium, magnesium, and manganese, cannot induce neural differentiation until the treated cells are transferred to a medium containing a sufficient concentration of sodium ions.[17] A solution containing 0.088 M NaCl induced differentiation into nerve and pigment cells, whereas a solution containing 0.044 M NaCl was ineffective. Sodium uptake in previously induced cells was twofold higher than in the uninduced cells.[17]

Aggregates from stage 10 and stage 11 gastrulae differentiate into functional muscle, mesenchyme, and nerve and pigment cells when the standard amount of sodium is present. In a reduced sodium concentration muscle and mesenchyme differentiate normally, but pigment cells are lacking and nerve cells are rare. Stage 12 aggregates, however, differentiate into nerve and pigment cells as well as functional muscle and mesenchyme when cultured in reduced sodium. The development of pigment cells may be as extensive in this medium as in normal sodium, but the nerve cells are always less extensive. Thus, induction in aggregates from stage 10 gastulae shows a strong sodium dependency, but during development, induction becomes increasingly independent of sodium.

Similarly, normal induction of nerve and pigment cells by

mesoderm in small explants prepared from the dorsal lip and lateral marginal zones of the early gastrula depends upon the external concentration of sodium. Nerve and pigment cells are induced by mesoderm when the culture medium contains 0.088 M NaCl. At a concentration of 0.044 M NaCl, the mesoderm differentiates into muscle and mesenchyme, but the induction of nerve and pigment cells does not occur.

The sodium dependency of induction in aggregates seemed at variance with the fact that induction in the whole gastrula occurs in very low concentrations of sodium and other ions, and in glass-distilled water as well. If ions have a role in induction in the whole egg, they must come from an endogenous source.[17] It has been shown[18] that the cations $Li^{2+}$ or $Ca^{2+}$ induce differentiation of neurons from presumptive epidermis. These neurons can be further induced to differentiate into pigment cells by subsequent treatment with the same ions. Three types of neurons or pigment cells are produced by sequential inductions. The ectodermal cells are competent for these inductions only at the stages of development at which the inductions normally occur in the embryo. The effect of an ionic inducer also changes with time, whether the cells are left in the embryo or cultured *in vitro*. Ectodermal cells cultured *in vitro* must be treated with an ion that can trigger them to become neurons before they are competent to be induced to become pigment cells. Therefore, it has been suggested[19] that changes in ionic flux normally accompany the induction of neural tissue *in vivo*. These changes do not occur in an interspecies hybrid that does not gastrulate normally and in which the ectoderm is not exposed to the natural inducer. The neural plate and its derivative cells are not induced, and the normal changes in permeability to ions do not occur. Cells from these abnormal gastrulae can be induced to form neurons and pigment cells by treating them with the same ions which induce neural differentiation in the cells of a normal embryo. This shows that the hybrid cells contain the information to make neurons and pigment cells and can express this information if appropriately triggered. These experiments further suggest that the ionic changes accompanying induction *in vivo* result from the process of induction and are sufficient for the induction of neurons and pigment cells.

## Interrelationships of Cations and Cyclic Nucleotides

Lithium is known to inhibit adenylate cyclase activity in a variety of organs, including brain *in vitro* (review [19-21]), but it is not known if it affects the intracellular level of cAMP. Lithium ions may also inhibit the effect of cAMP. A small amount of bound calcium is required for the activation of adenylate cyclase from brain and for at least one protein kinase.[19] However, concentrations ($\leqslant 10^{-4}$ M) of calcium inhibit adenylate cyclase and protein kinase from brain and other tissues, and even cause cAMP to inhibit cAMP-dependent protein kinase.[19] Cyclic AMP phosphodiesterase activity is also inhibited by calcium ($\leqslant 10^{-4}$ M), particularly in the absence of magnesium, although smaller amounts of calcium ($\geqslant 10^{-4}$ M) may stimulate it.[19] Calcium also inhibits the activity of phosphoprotein phosphatase.[19] Magnesium and manganese ions are required for the activity of adenylate cyclase, phosphodiesterase, and protein kinases. Magnesium also affects the modulation of protein kinase activity by the cAMP-dependent protein kinase modulator, but it is not required for the activity of at least one phosphoprotein phosphatase.[19] Manganese stimulates the activity of brain protein phosphatase.[19] Zinc can inhibit adenylate cyclase, phosphodiesterase, and protein phosphatase; but it neither increases the activities of protein kinases nor appears to inhibit them.[19] Sodium and potassium ions have only a slight effect on adenylate cyclase. A high concentration of either sodium or potassium inhibits the ability of protein kinases to phosphorylate their substrates.[19] For example, 0.1 M NaCl increases the Michaelis constant for casein substrate by six times; however, a similar concentration may stimulate phosphorylation of histone substrate. High concentrations of potassium also increase the cAMP level in cerebral cortex.[22] Thus, the relationships between cations and cyclic nucleotides are complex. Many studies have been conducted at abnormally high ion concentrations, and may therefore have little relevance to physiological conditions. It is known that an elevation of cAMP causes changes in the levels of intracellular ion concentrations. Hence, changes in the level of cyclic nucleotide may cause alterations in ionic environment that can modify gene expression.

## CURRENT STATUS OF INDUCTION MECHANISMS OF NEURAL TISSUE

Although Sutherland and his colleagues established in 1960 that cAMP may act as a second messenger for the action of several hormones, not until 1970 was it discovered that cAMP may be an important factor in regulating the expression of many differentiated functions in mouse neuroblastoma cells in culture.[6-8] One can summarize the normal induction mechanism for neural tissue in the following manner. A specific polypeptide may increase the level of cAMP in the ectoblast, which then initiates the formation of neural tissue. A further change in the cAMP level may occur as a result of the effect of maternal neurotransmitters, since some of the neurotransmitters are known to increase the cAMP level in nerve cells.[8, 23, 24] An elevated cAMP level is known to increase the level of cAMP binding protein[25] as well as cAMP phosphodiesterase in nerve cells.[26] The increase in the level of cAMP binding protein may be one of the important intracellular mechanisms responsible for maintaining high cAMP levels during further differentiation, since the protein-bound cAMP is resistant to enzymatic hydrolysis. Therefore, the increased level of cAMP phosphodiesterase activity does not significantly affect the continued increase in the cAMP level. cAMP in some way inhibits the expression of certain genes, while it increases the expression of others.[27] One of the mechanisms of cAMP action involves the phosphorylation of specific proteins for its physiological effect. Therefore, an increase in cAMP-dependent phosphorylation activity during differentiation can be predicted. Indeed, using neuroblastoma cell culture as an experimental model, we have shown[28] that the cAMP-dependent phosphorylation of specific proteins increases in the cytosol. However, cAMP also initiates changes responsible for marked changes in cAMP-independent phosphorylation.[28] For example, during differentiation of neuroblastoma cells cAMP-independent phosphorylation increases in the nuclear fraction, but decreases in the cytosol fraction. On the other hand, the overall phosphorylation activity of the pellet fraction, which contains ribosomes, mitochondria, and membranous structures, was decreased during differentiation.[28] Thus, cAMP may be involved in the mechanism of induction of neural tissue from ectoderm. Further differentiation and maturation require the inhibition of cell division. Therefore, cAMP in some way must turn off cell division at a precise time during development. Indeed, it has been observed[29] that the synthesis of histone and phosphorylation of $H_1$-histone, which are linked with cell proliferation,[30-33] are markedly decreased in cAMP-induced "differentiated" neuroblastoma cells that have stopped cell division. In contrast to histone synthesis and $H_1$-histone phosphorylation, no significant change occurs in either the synthesis or phosphorylation of nonhistone chromosomal proteins in cAMP-induced differentiated neuroblastoma cells. Other investigators[34-36] have been unable to demonstrate any significant change in nonhistone chromosomal proteins during development of the rat brain. The significance of a small decrease in the synthesis and the small increase in the phosphorylation of a 40,000-dalton peptide in cAMP-induced differentiated cells is unknown.[29] It has been suggested[37] that a nonhistone protein in the 40,000 to 45,000-dalton range may be involved in DNA replication in eukaryotic cells. Since DNA synthesis in differentiated neuroblastoma cells is decreased by about 90%,[38] the change in the synthesis and phosphorylation of the 40,000-dalton peptide may be related to the inhibition of DNA synthesis. cGMP is not involved in the mechanism of neural induction[9] or in further differentiation.[8]

After neural induction, what factor initiates the separation of glial and nerve cells and various types of nerve cell? The answer to this fundamental question remains unknown.

## REFERENCES

1. Spemann, H. *Experimentelle Beiträge Zu einen Theorie der Entwicklung*, Verlag Von Julius, Springer, Berlin, 1936.
2. Saxen, L., and Toivonen, S., *Primary Embryonic Induction*, Elek Books/Logos Press, London, 1962.
3. Brachet, J., *Chemical Embryology* (trans. by L. G. Barth), 523 pp. Hafner Press, New York, 1968.
4. Brachet, J., *Introduction to Molecular Embryology*, 170 pp., Springer-Verlag, New York, 1974.
5. Tiedeman, H., Biochemical aspects of primary induction and determination, in: *Biochemistry of Animal Development*, Vol. 2 (R. Weber, ed.), pp. 4–55, Academic Press, New York, 1967.
6. Prasad, K. N., and Hsie, A. W., Morphological differentiation of mouse neuroblastoma cells induced *in vitro* by dibutyryl adenosine 3':5'-cyclic monophosphate, *Nature (London), New Biol.* **233**:141–142, 1971.
7. Furmanski, P., Silverman, D. J., and Lubin M., Expression of differentiated functions in mouse neuroblastoma mediated by dibutyryl-cyclic adenosine monophosphate, *Nature (London)* **233**:413–415, 1971.
8. Prasad, K. N., Differentiation of neuroblastoma cells in Culture, *Biol. Rev.* **50**:129–165, 1975.
9. Wahn, H. L., Lightbody, L. E., Tchen, T. T., Induction of neural differentiation in culture of amphibian undetermined presumptive epidermis by cyclic AMP derivatives, *Science* **188**:366–369, 1975.
10. Prasad, K. N., Role of cyclic AMP in the differentiation of neuroblastoma cell culture, in: *The Role of Cyclic Nucleotides in Carcinogenesis*, Vol. 6 (J. Schultz and H. G. Gratzner, eds.), pp. 207–237, Academic Press, New York, 1973.
11. Wahn, H., Lightbody, L. T., Tchen, T. T., and Taylor, J. D., Adenosine 3',5'-monophosphate, morphogenetic movements and embryonic neural differentiation in *Pleurodeles waltili*, *J. Exp. Zool.* **196**:125–130, 1976.
12. Wahn, H. L., Taylor, J. D., and Tchen, T. T., Acceleration of amphibian embryonic melanophore development by melanophore-stimulating hormone, $N^6O^2$-dibutyryl adenosine 3',5'-monophosphate and theophylline, *Dev. Biol.* **49**:470–478, 1976.

13. Weston, J. A. The migration and differentiation of neural crest cells, in: *Advances in Morphogenesis*, Vol. 8 (M. Abercrombie and J. Brachet, eds.); pp. 41–114, Academic Press, New York, 1970.

14. Yamada, T., A chemical approach to the problem of the organizer, in: *Advances in Morphogenesis*, Vol. 1 (M. Abercrombie and J. Brachet, eds.), pp. 1–54, Academic Press, New York, 1961.

15. Reporter, M., and Rosenquist, G. C., Adenosine 3',5'-monophosphate: Regional differences in chick embryos at the head process stage, *Science* **178**:628–630, 1972.

16. Barth, L. G., and Barth, L. J., Differentiation of cells in the *Rana pipiens* gastrula in unconditioned medium, *J. Embryol. Exp. Morphol.* **7**:210–222, 1959.

17. Barth, L. G., and Barth, L. J. The sodium dependence of embryonic induction, *Dev. Biol.* **20**:236–262, 1969.

18. Barth, L. G., and Barth, L. J., Sodium and calcium uptake during embryonic induction in *Rana pipiens*, *Dev. Biol.* **28**:18–34, 1972.

19. McMahon, D., Chemical messenger in development: A hypothesis, *Science* **185**:1012–1021, 1974.

20. Forn, J. and Valdecasas, F. G., Effect of lithium on brain adenyl cyclase activity, *Biochem. Pharmacol.* **20**:2773–2779, 1971.

21. Smith, B. M., Harris, C. A., and Major, P. W., The effect of lithium ions on the activation of ovarian adenyl cyclase, in: *Advances in Cyclic Nucleotide Research*, Vol. 1 (P. Greengard and G. A. Robison, eds.), p. 588(abstr.), Raven Press, New York, 1972.

22. Shimizu, H., Creveling, C. R., and Daly, J. W., Effect of membrane depolarization and biogenic amines on formation of cyclic AMP in incubated brain slices, in: *Role of Cyclic AMP in Cell Function*, Vol. 3 (P. Greengard and E. Costa, eds.), pp. 135–154, Raven Press, New York, 1972.

23. Gilman, A. G., and Nirenberg, M., Regulation of adenosine 3'-5'-cyclic monophosphate metabolism in cultured neuroblastoma cells, *Nature (London)* **234**:356–357, 1971.

24. Blume, A. J., Dalton, C., and Sheppard, H., Adenosine-mediated elevation of cyclic 3'-5'-adenosine monophosphate concentrations in cultured mouse neuroblastoma cells, *Proc. Natl. Acad. Sci. U.S.A.* **70**:3099–3102, 1972.

25. Prasad, K. N., Sinha, P. K., Sahu, S. K., and Brown, J. L., Binding of cyclic nucleotides with soluble proteins increases in "differentiated" neuroblastoma cells in culture, *Biochem. Biophys. Res. Commun.* **66**:131–138, 1975.

26. Prasad, K. N., and Kumar, S., Cyclic 3',5'-AMP phosphodiesterase activity during cyclic AMP-induced differentiation of neuroblastoma cells in culture, *Proc. Soc. Exp. Biol. Med.* **142**:406–209, 1973.

27. Prasad, K. N., Sahu, S. K., and Sinha, P. K., Cyclic nucleotides in the regulation of expression of differentiated functions in neuroblastoma cells, *J. Natl. Cancer Inst.* **57**:619–632, 1976.

28. Ehrlich, Y. H., Brunngraber, E. G., Sinha, P. K., and Prasad, K. N., Specific alterations in phosphorylation of cytosol proteins from differentiating neuroblastoma cells grown in culture, *Nature (London)* **265**:238–240, 1977.

29. Lazo, J. S., Prasad, K. N., and Ruddon, R. W., Synthesis and phosphorylation of chromatin-associated proteins in cAMP-induced "differentiated" neuroblastoma cells in culture, *Exp. Cell Res.* **100**:41–46, 1976.

30. Balhorn, R., Bordwell, J., Sellers, L., Granner, D., and Chalkley, R., Histone phosphorylation and DNA synthesis are linked in synchronous cultures of HTC cells, *Biochem. Biophys. Res. Commun.* **46**:1326–1333, 1972.

31. Gurley, L. R., Walters, R. A., and Tobey, R. A., The metabolism of histone fractions. IV. Synthesis of histones during the $G_1$-phase of the mammalian life cycle, *Arch. Biochem. Biophys.* **148**:633–641, 1972.
32. Gurley, L. R., Walters, R. A., and Tobey, R. A., Cell cycle-specific changes in histone phosphorylation associated with cell proliferation and chromosome condensation, *J. Cell Biol.* **60**:356–364, 1974.
33. Krause, M. O., and Inasi, B. S., Histones from exponential and stationary L-cells. Evidence for metabolic heterogeneity of histone fractions retained after isolation of nuclei, *Arch. Biochem. Biophys.* **164**:179–184, 1974.
34. Burdman, J. A., The relationship between DNA synthesis and the synthesis of nuclear proteins in rat brain during development, *J. Neurochem.* **19**:1459–1469, 1972.
35. Fujitani, H., and Holoubek, V., Nonhistone nuclear proteins of rat brain, *J. Neurochem.* **23**:1215–1224, 1974.
36. Olpe, H. R., Van Hahn, H. P., and Honegger, C. G., The non-histone protein pattern of rat brain during ontogenesis, *Experientia* **29**:665–666, 1972.
37. Elgin, S. C. R., Boyd, J. B., Hood, L. E., Wray, W., and Wu, F. C., A prologue to the study of the nonhistone chromosomal proteins, *Cold Spring Harbor Symp. Quant. Biol.* **38**:821–833, 1973.
38. Prasad, K. N., Waymire, J. C., and Weiner, N., A further study on the morphology and biochemistry of x-ray and dibutyryl cyclic AMP-induced differentiated neuroblastoma cells in culture, *Exp. Cell Res.* **74**:110–114, 1972.

# ROLE OF CYCLIC NUCLEOTIDES IN REGULATION OF DIFFERENTIATION OF NERVE CELLS

## INTRODUCTION

The possible involvement of adenosine 3′,5′-cyclic monophosphate (cAMP) in neural differentiation became apparent when it was demonstrated that an elevation of the intracellular level of cAMP in neuroblastoma cells induces, as well as increases, the expression of many differentiated functions characteristic of mature neurons.[1] The process of neural differentiation involves many steps, including induction, cell migration, regulation of induced differentiated functions, and inhibition of cell division. However, neuroblastoma cells in many ways are differentiated already and possess several features of mature neurons which are expressed mostly at low levels. Therefore, neuroblastoma cells in culture may be suitable primarily for studying the regulation of differentiated functions, which are induced already, and in identifying genetic and structural features modified experimentally. To study the involvement of cAMP in neural induction, the experimental system developed by early embryologists or embryoid cells of teratocarcinoma must be used. Indeed, by using explants of gastrulae of amphibia, it has been shown[2] that cAMP induces neural differentiation. An extensive modification of gene expression, associated with structural organization, must occur during the period of differentiation. Genetic and structural modifications

occur sequentially, with each differentiated function becoming detectable at a precise time and for a defined purpose. The purpose of
this chapter is to discuss the role of cAMP in regulating differentiated
functions of nerve cells.

## EXPERIMENTAL MODELS

The modification of gene expression by cAMP has been studied
in neuronal tumor cell lines derived from mouse (transplanted tumor
originally occurring spontaneously in the body cavity), rat (chemically
induced tumor in central nervous system), and human neoplasms
(tumor cells obtained with the bone marrow or from abdominal tumor
mass). In addition, it has been studied in explants of normal embryonic nervous tissue and in hybrids of neuroblastoma and glioma
cells, in which the expression of certain genes is amplified, and the
expression of others is decreased and/or extinguished. The clonal cell
lines with diploid chromosome number may provide a simple experimental system to study the regulation of gene expression in
nerve cells. The mouse neuroblastoma cells are highly aneuploid; the
chromosome number varies from 59 to 192 (the diploid number is 40).
Human neuroblastoma cells[4] and rat central nervous system tumor
cells[5] are nearly diploid. Therefore, one has to be cautious in evaluat-
ing the modification of gene expressions in various cell lines during
differentiation.

## METHODS OF EVALUATING CHANGES IN GENETIC AND STRUCTURAL EXPRESSIONS

Gene expression may be measured by assaying the activity
and/or amount of a direct gene product. An elevation of the intracellular level of cAMP in certain clones of neuroblastoma cells increases or
decreases the expression of several genes. The modification of genetic
expression by cAMP may occur at the translational or transcriptional
level or during posttranslational events. The interpretation of the effect on transcriptional or translational level has been primarily based
on the use of metabolic inhibitors. If the cAMP-induced increase in
the activity of a gene product can be blocked by actinomycin D (inhibits mRNA synthesis), the effect is considered the result of changes at
the transcriptional level. On the other hand, if cAMP-induced increase
in activity of a gene product is inhibited by cycloheximide (inhibits

protein synthesis) but not by actinomycin D, the effect is considered the result of alterations at the translational level. However, these metabolic inhibitors may produce effects other than inhibition of protein or RNA synthesis on the intact cells in culture; therefore, one has to be cautious in interpreting such data.

Structural alterations in cells have been measured by several methods. Cytoplasmic processes ($>50$ $\mu$m in length) showing electrophysiological responses have been referred to as dendritic, neuritic, or axonlike processes. The rearrangement of microtubules and microfilaments and changes in other cytoplasmic and nuclear structures during differentiation have been demonstrated by electron microscopy. Vinblastine sulfate (inhibits assembly of microtubules) and cytochalasin B (inhibits assembly of microfilaments) have been used to demonstrate the role of these cytoplasmic structures in the expression and maintenance of differentiated phenotype. These drugs are also known to affect other metabolic functions; therefore, one has to be careful in interpreting the effect of these agents on the basis of their effects on filamentous structures alone.

Changes in gene expression, including morphological characteristics, may be modified by agents which do not increase the cAMP level. This observation does not necessarily negate the role of cAMP in the regulation of genetic and structural features. The role of cAMP in the regulation of cellular function should not be evaluated by measuring only changes in the intracellular level of cAMP. One of the postulated mechanisms of cAMP effects on mammalian cells is that changes in cAMP-dependent phosphorylation precede any change in the expression of a cAMP-mediated function. Since the binding of cAMP with the regulatory subunit of cAMP-dependent protein kinase is essential for the dissociation of the catalytic subunit to act in the phosphorylation of proteins, a change in the intracellular level of cAMP would determine the amount of catalytic subunit of cAMP-dependent protein kinase available for phosphorylation. It is possible that cAMP-dependent functions can be altered without any change in the cAMP level. For example, alteration of cAMP-dependent phosphorylation activity can be achieved by changing the phosphatase activity at the phosphorylation site of the phosphoprotein without an alteration in cellular cAMP. Some agents that mimic the effect of cAMP without actually changing the level of cAMP may act by directly affecting the phosphatase activity. The involvement of cAMP in the regulation of a specific gene product or morphological alteration is considered plausible only when the following criteria are met: (1) The effect must be produced by at least one analog of cAMP, one stimulator of adenylate cyclase, and one inhibitor of cyclic nucleotide phosphodiesterase; (2) both the stimulator of adenylate cyclase and the inhibitor of phosphodiesterase must increase the intracellular level of cAMP; (3) changes in cAMP-dependent phosphorylation must precede alteration in cellular functions. The postulated mechanism of cAMP effects on eukaryotic cells involves changes only in cAMP-dependent phosphorylation activity; in prokaryotes it also involves activation of certain gene expressions by cAMP-binding proteins. The latter mechanism has not been demonstrated in the mammalian cells. cAMP binding proteins distinct from the regulatory subunit have been demonstrated in at least two separate studies.[6,7] The possibility that cAMP effects involve mechanisms other than phosphorylation should be noted.

## MODIFICATION OF GENE EXPRESSION AND MORPHOLOGY UNIQUE TO CELL CULTURE CONDITION AND EXPERIMENTAL MODELS

Phenotypic expression changes when neuroblastoma cells are transferred to a culture condition.[8] This may indicate a transition in regulating mechanisms controlling gene expression concomitant with environmental modification. For example, the electrophoretic band of 6-phosphogluconate dehydrogenase is very distinct in mouse neuroblastoma cells grown *in vivo*, but this band is barely detectable in cells grown in culture (Figure 3-1). The reverse is true for glucose-6-phosphate dehydrogenase (Figure 3-2). The activities of fructose diphosphate aldolase, glucose-6-phosphate dehydrogenase, lactic dehydrogenase, glutamate dehydrogenase, and malate dehydrogenase in neuroblastoma cells in culture are lower than those in freshly isolated neuroblastoma cells.[9] The mouse neuroblastoma cells grown *in vivo* contain ganglioside $GT_{1a}$; however, this ganglioside is undetectable in cells grown in culture.[10] A twofold increase in tyrosine hydroxylase (TH) activity is observed when the clone is developed to grow in culture[11] instead of *in vivo*. In contrast, we found[12] that TH activity markedly decreased when uncloned neuroblastoma cells were grown in culture (Table 3-1). The enzyme activity of freshly isolated cells dropped from 58 to 19 pmol/min per $10^6$ cells after 3 days in culture. The enzyme activity dropped further to a 2 pmol value at 100 days and then appeared to stabilize. This may be due to selecting cells in culture with a low activity of TH. However, a marked decrease in enzyme activity after 3 days in culture suggests that regulatory

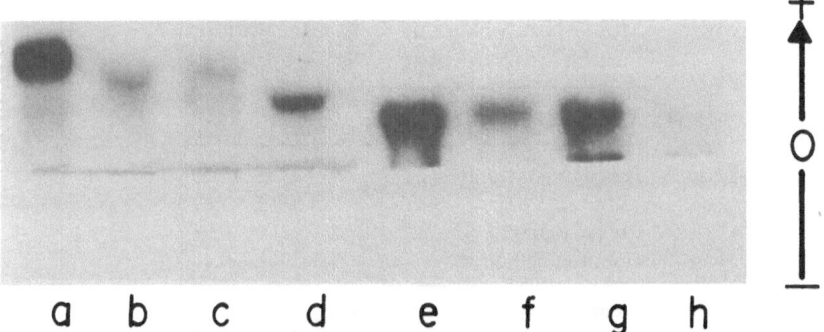

**Figure 3-1.** Starch gel electrophoretic pattern of 6-phosphogluconate dehydrogenase. O, origin; a, mouse neuroblastoma cells grown *in vivo*; b, control mouse neuroblastoma cells in culture; c, mouse neuroblastoma cells treated with Ro 620-1724, a phosphodiesterase inhibitor; d, mouse neuroblastoma cells treated with prostaglandin $E_1$ ($PGE_1$); e, papaverine-treated human neuroblastoma cells; f, human neuroblastoma cells treated with sodium butyrate; g, control human blastoma cells; h, human neuroblastoma cells treated with $PGE_1$. (From Prasad *et al.*[8])

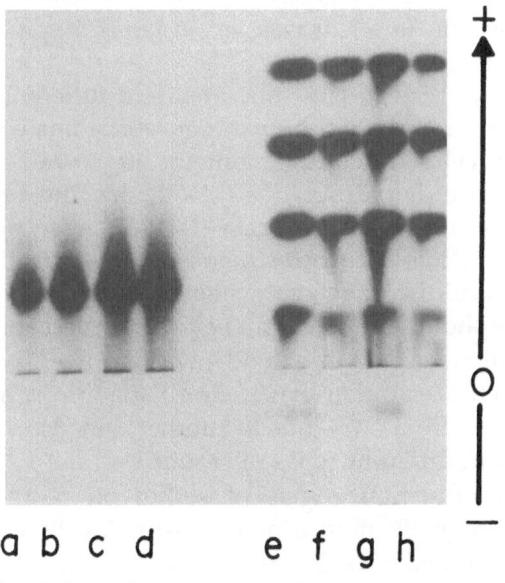

**Figure 3-2.** Starch gel electrophoretic pattern of glucose 6-phosphate dehydrogenase. O, origin; a, mouse neuroblastoma cells treated with $PGE_1$; b, mouse neuroblastoma cells treated with Ro 20-1724; c, control mouse neuroblastoma cells in culture; d, mouse neuroblastoma cells grown *in vivo*; e, papaverine-treated human neuroblastoma cells; f, human neuroblastoma cells treated with sodium butyrate; g, control human neuroblastoma cells; h, human neuroblastoma cells treated with $PGE_1$. (From Prasad *et al.*[8])

**Table 3-1.** TH Activity in Uncloned
Mouse Neuroblastoma Cells after Various
Periods of Culturing[a]

| Condition | TH activity[b,c] |
|---|---|
| Fresh grated tumor cells | 57.93 ± 15.20 |
| Cultured 3 days[d] | 18.50 ± 4.62 |
| Cultured 100 days[d] | 1.42 ± 0.52 |
| Cultured 380 days[d] | 1.38 ± 0.48 |

[a] From Waymire et al.[12]
[b] Picomoles of $^{14}CO_2$ in 30 min per $10^6$ cells.
[c] Each value is the average of four separate determinations ± SE.
[d] Medium was changed every second day.

changes associated with genetic expression of TH were modified by growth conditions, especially because a loss of cells was not detected during this period. However, it is possible that once enzyme activity of a given clone is stabilized in culture the activity would not change further, regardless of whether the cells are grown *in vivo* or *in vitro*. Additional gene products whose expression may change as a function of growth condition (e.g., *in vivo* vs. *in vitro*) may occur, but they have yet to be identified.

We have observed at least one apparent difference in gene expression between mouse and human neuroblastoma cell line. Mouse neuroblastoma cells in culture contain only one electrophoretic band[13,14] of lactate dehydrogenase (LDH), the fifth band of muscle type; and human neuroblastoma cells (IMR-32) have all five bands of LDH (Figure 3). Mouse neuroblastoma cells *in vivo* exhibit round-cell morphology. In culture, most of the neuroblastoma cells maintain this round-cell morphology; however, they vary noticeably in size and shape and in their ability to extend neurites. Even within a single clonal line, such a variation exists. The variation in morphology is observed not only from one clone to another, but also when the same clone is grown in different types of serum.[15]

The effect of various types of serum on morphological and biochemical changes in mouse neuroblastoma cells (clone $NBP_2$) in culture was studied (Table 3-2). The extent of spontaneous morphological differentiation varied markedly depending upon the type of serum, and was maximal in agammaglobulin calf serum (CS). The extent of morphological treatment of cells with cAMP-stimulating agents also depended upon serum type, and was least pronounced in fetal calf serum. The doubling time and extent of clumping varied

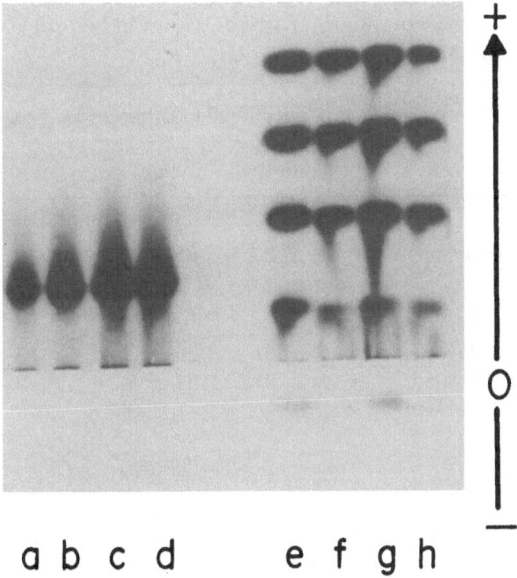

**Figure 3-3.** Starch gel electrophoretic pattern of lactate dehydrogenase. O, origin; a, control mouse neuroblastoma cells in culture; b, mouse neuroblastoma cells treated with Ro 20-1724; c, mouse neuroblastoma cells treated with PGE$_1$; d, mouse neuroblastoma grown *in vivo*; e, papaverine-treated human neuroblastoma cells; f, human neuroblastoma cells treated with sodium butyrate; g, control human neuroblastoma cells; and h, human neuroblastoma cells treated with PGE$_1$. (From Prasad *et al.*[8])

with the type of serum. The activity of TH in NB cells depended upon serum type, and was highest in newborn CS and agammaglobulin CS. Although elevation of intracellular levels of cAMP in NBP$_2$ clone invariably stimulates neurite formation and TH activity, these functions were increased in certain sera without a significant increase in the cellular cAMP levels. The study shows that neurite formation, growth rate, and TH activity are regulated by more than one mode, one of which is is mediated by cAMP. The above changes are independently regulated in the sense that the expression of one can be increased in the absence of others.

It should be noted that the concentration requirement of each drug for a maximal effect on morphological differentiation also varies as a function of type of serum. Based on the results with serum type, we suggest that changes in serum content may be responsible in part for the spontaneous differentiation known to occur in a certain percentage of neuroblastoma cells *in vivo*, recognizing that all variants of neuroblastoma cells in any given tumor mass would not necessarily respond in a similar manner. Finally, the results also emphasize the

**Table 3-2.** Morphological Differentiation of Neuroblastoma
Cells (NBP$_2$ Clone) in Various Sera

| Type of serum | Percentage of differentiated cells[a,b] | | |
|---|---|---|---|
| | Control | PGE$_1$ | Ro 20-1724 |
| Fetal calf serum | 11 ± 2* | 14 ± 2† | 26 ± 3 |
| Heat-inactivated | 2 ± 0 | 81 ± 5†† | 51 ± 4[∂] |
| Dialyzed | 6 ± 1 | 72 ± 3** | 63 ± 4 |
| Agammaglobulin | ———— Toxic, poor growth ———— | | |
| Newborn calf serum | 2 ± 0 | 20 ± 2 | 60 ± 4 |
| Heat-inactivated | 5 ± 0 | 34 ± 3 | 40 ± 3 |
| Agammaglobulin | 2 ± 0 | 41 ± 3 | 75 ± 4 |
| Calf serum | ————Toxic, complete lethality ———— | | |
| Heat-inactivated | 66 ± 5 | — | — |
| Dialyzed | 28 ± 3 | 56 ± 4 | 93 ± 3 |
| Agammaglobulin | 77 ± 3 | 92 ± 2 | 89 ± 3 |

[a] Cells (25,000) were plated in 60-mm Falcon plastic dishes and prostaglandin E$_1$ (PGE$_1$) and 4-(3-butoxy-4-methoxybenzyl)-2-imidazolidinone (Ro
20-1724) were added separately 24 hr after plating. The drug and medium
were changed at days 2 and 3, and the percent of morphologically differentiated cells (processes >50 μm in length) was determined at 4 days after
treatment. All PGE$_1$ concentrations were 10 μg/ml except *, and all Ro 20-
1724 concentrations were 100 μg/ml except †. These were optimal concu-
bations which did not cause cell death. Each value represents the mean six
samples. From Prasad.[15]
[b] Symbols: *, standard deviation; †, 5 μg/ml; ††, 15 μg/ml; **, 2 μg/ml; [∂], 200
μg/ml.

hazard of comparing measurements from separate studies, using
identical clones, but culturing the cells in the presence of different
sera.

## REGULATION OF MORPHOLOGICAL DIFFERENTIATION

### Formation of Dendrites and Axons

A mature normal neuron contains well-defined dendritic processes and axons. It is possible that no two neurons have exactly the
same dendritic and axonal patterns, but nerve cells of the same type
tend to have similar, and frequently distinctive, types of outgrowth.
For example, retina have many bipolar neurons, whereas mammalian
Purkinje neurons have an elaborate configuration of dendritic pro-

cesses. The development of distinctive forms must be largely due to the interactions of developing cells with neighbors, but intrinsic factors may also play a role.[16] Properties such as the number, size, and symmetry of initial outgrowths, their rate of growth, and frequency of branching could be largely independent of external influences and may contribute to the final form of neurons. Indeed, it was found[17] that the final shape of a neuron may result from autonomous activities of the growth cone. Other parts of the cell may have played a supportive role, but had no obvious influence on the final pattern of branches formed. When a single neuron was observed in culture, it was found that the tips of the fibers advance in straight lines and grow at rates that do not vary appreciably with time. Most of the branch points are formed by the bifurcation of the growth cone, apparently at random, and thereafter remained at about the same distance from the cell body (Figure 3-4). In spite of elegant descriptions of the morphology and growth pattern of dendritic processes, very little is known about factors responsible for the induction of these morphological processes, and biochemical steps involved in the expression of the differentiated phenotype. Using mouse and human neuroblastoma cells, rat CNS neural tumor cells, and embryonic nervous tissue, several agents which induce morphological differentiation have been identified.

*Regulation of Neurite Formation*

The cytoplasmic processes in neuroblastoma cells ($>50$ $\mu$m in length) have been arbitrarily referred to as neurites or axonlike processes and are considered an expression of morphological differentiation. Although these processes are electrically excitable and are capable of generating action potentials,[18] cells having such processes may not express many biochemically differentiated functions.[1] Spontaneous neurite formation occurs in both mouse and human neuroblastoma cells in culture. Many agents, such as dibutyryl cAMP,[20,21] PGE$_1$,[22] inhibitors of phosphodiesterase,[23] serum-free medium,[24] X-irradiation,[25] 5-bromodeoxyuridine,[26,27] trifluoro-methyl-2-deoxyuridine,[28] 6-thioguanine,[29] cytosine arabinoside,[3] methotrexate,[31] glial extract,[32,33] hypertonic medium,[34] nerve growth factor,[35,36] and dimethylsulfoxide[37] induce neurite formation in neuroblastoma cells in culture. The formation of neurites after treatment of mouse and human neuroblastoma cells with various agents is shown in Figures 3-5A–D and 3-6A–D. Many of the agents which induce neurite formation increase the intracellular level of cAMP, whereas others do so

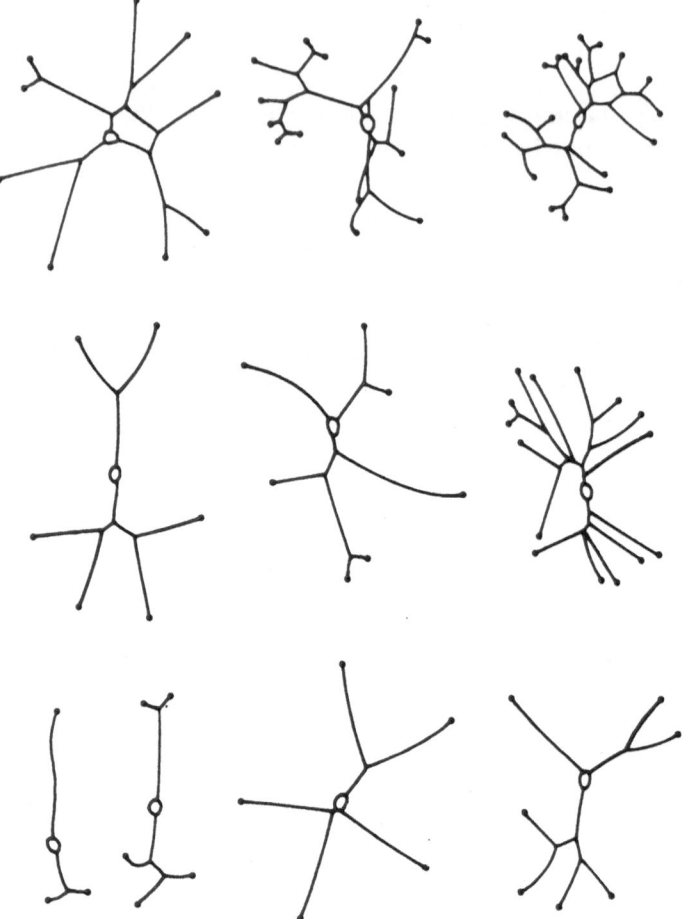

**Figure 3-4.** Outlines of a number of sympathetic neurons growing singly in culture. Bar, 100 μm. (From Bray.[17])

without changing the cAMP level.[1] These data suggest that neurite extension is regulated by at least two modes, one of which involves changes in the cAMP level.

Various agents induce neurites, probably by promoting the organization of microtubules and microfilaments,[1] since the expression of the morphologically differentiated phenotype is blocked by vinblastine sulfate (interferes with the assembly of microtubules) and cytochalasin B (interferes with the assembly of microfilaments). The organization of microtubules is more important than the organization of microfilaments for the maintenance of neurites.[38] Halothane, a commonly used volatile anesthetic, inhibits the expression of neurites

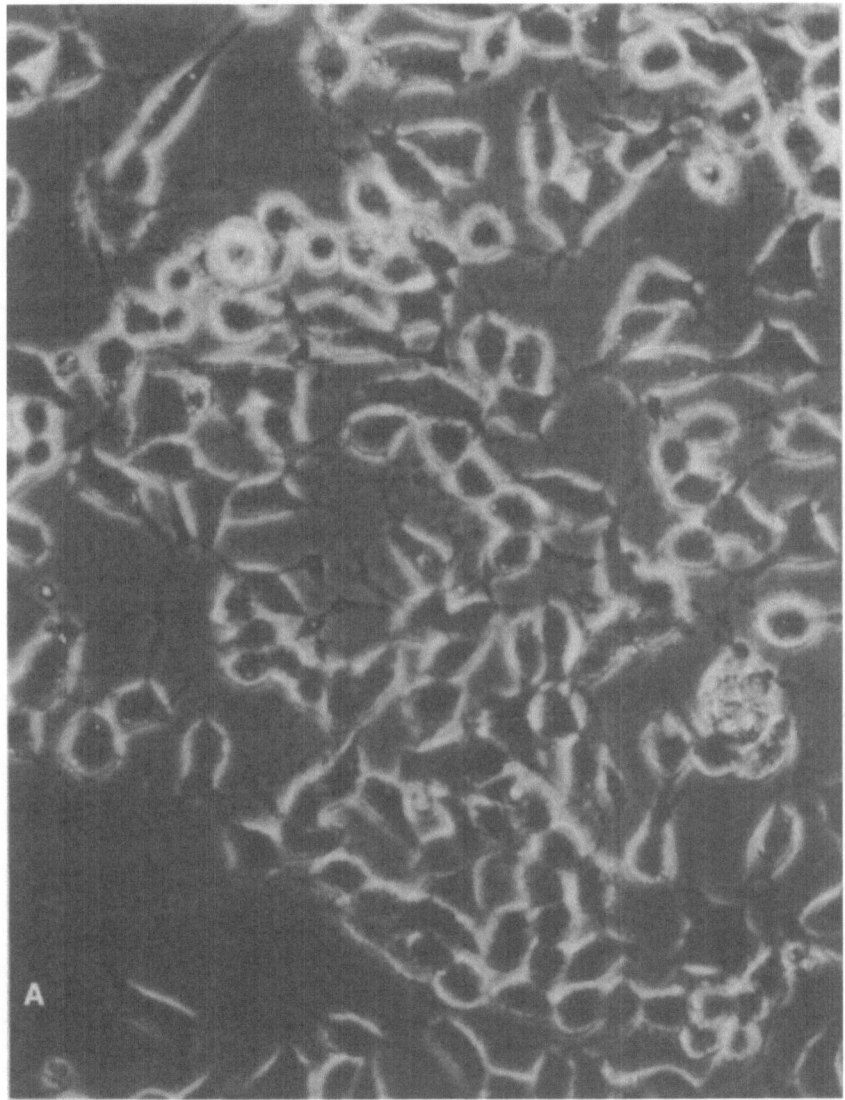

**Figure 3-5.** Phase contrast micrographs (through page 61) of mouse neuroblastoma cells (NBA$_{2(1)}$ clone) in culture. Cells (50,000) were plated in 60-mm Falcon plastic dishes, and PGE$_1$ (10 $\mu$g/ml) and Ro 20-1724 (200 $\mu$g/ml) were added separately 24 hr later. The medium and drug were changed at 3, 5, 8, and 11 days after treatment. Control culture (A) shows that cells grow in clumps and some of them have short cytoplasmic processes. PGE$_1$-treated culture 4 days after treatment (B) shows formation of long neurites. PGE$_1$-treated culture 14 days after treatment (C) shows that the remaining cells maintain their differentiated phenotype and show extensive network of thickened fibers. (From Prasad and Kumar.[19])

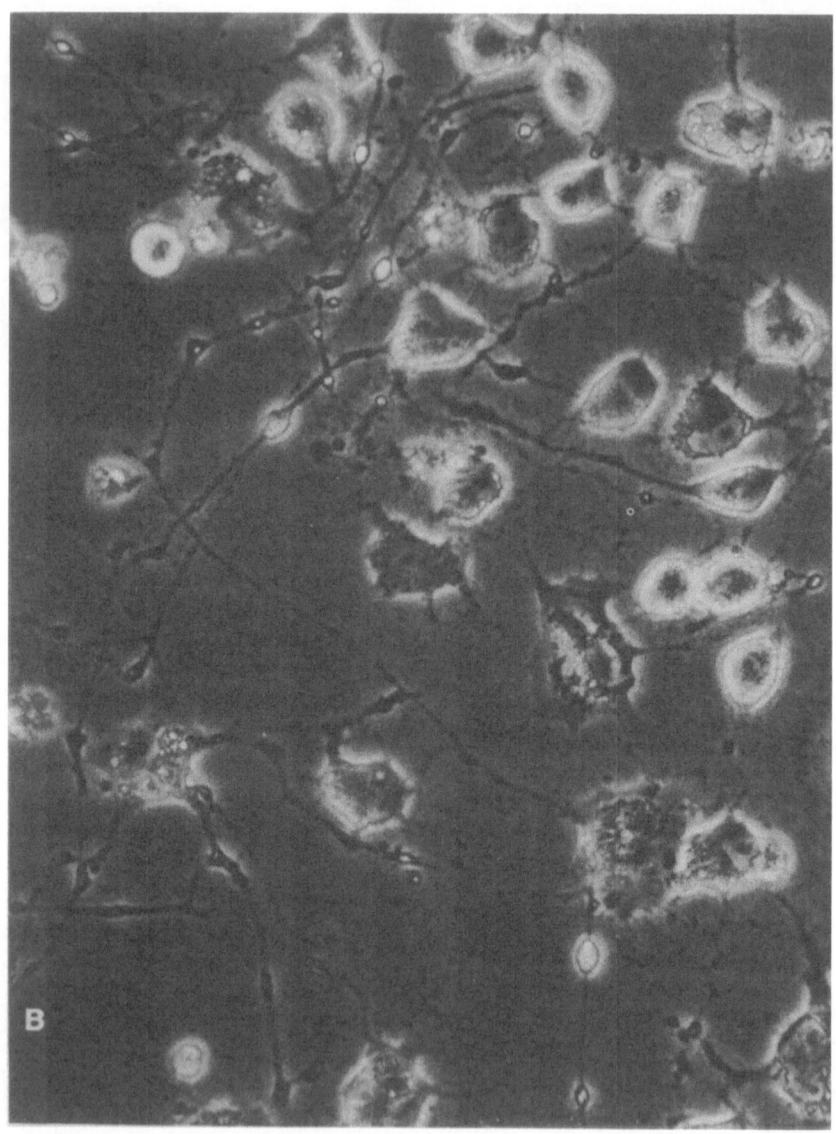

Figure 3-5B

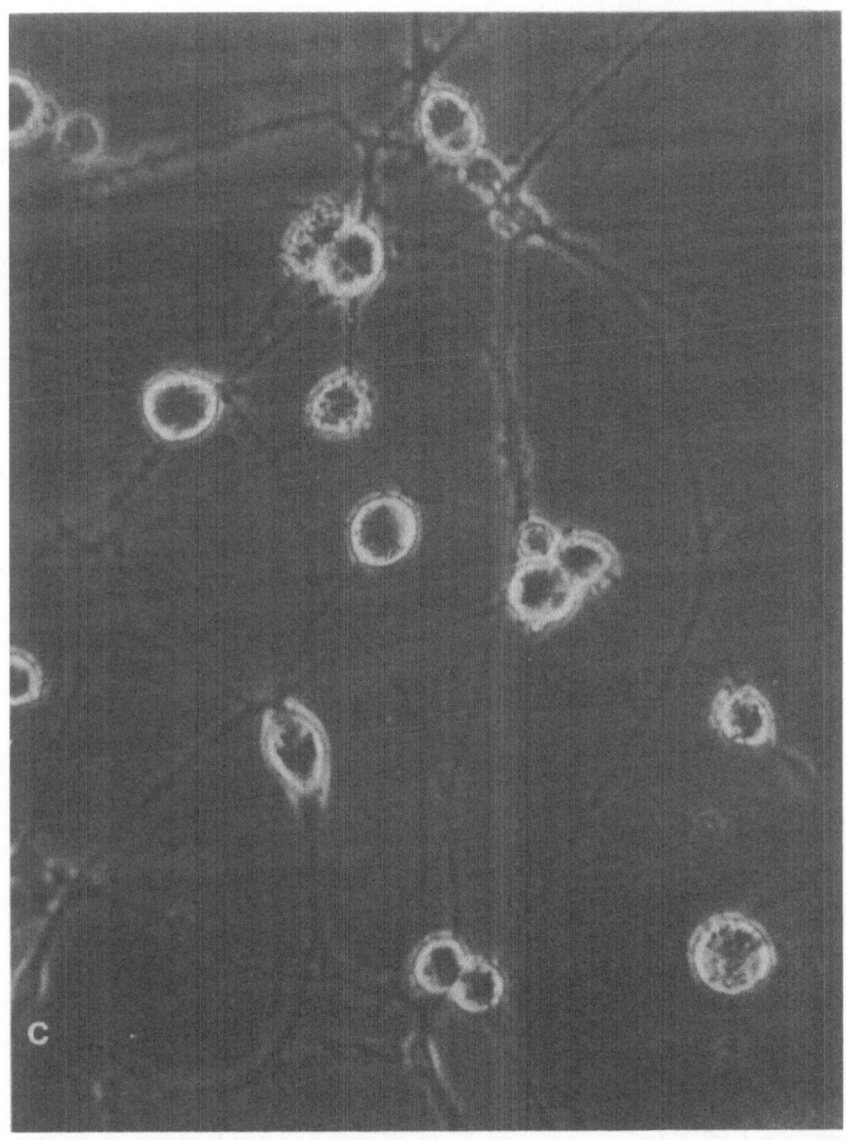

**Figure 3-5C**

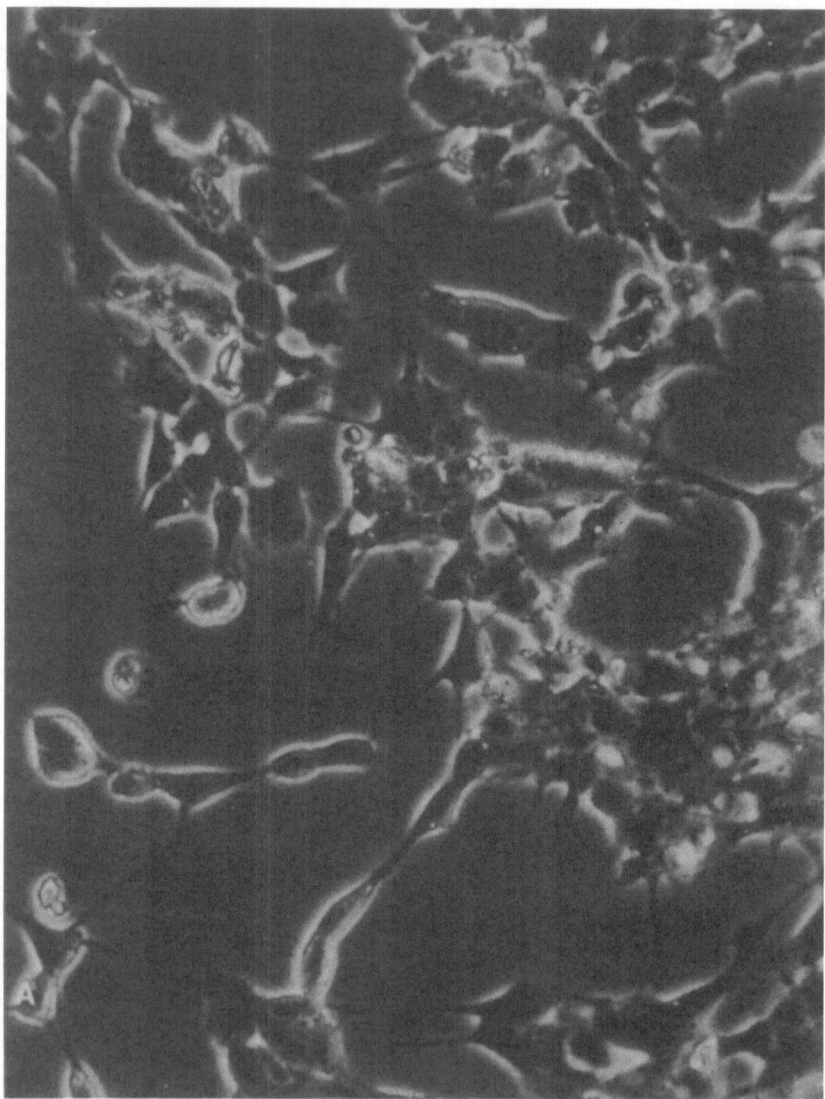

**Figure 3-6.** Phase-contrast micrographs (through page 66) of human neuroblastoma cells in culture (IMR-32 line). Cells were plated in 60-mm Falcon plastic dishes and papaverine (2.5 μg/ml), sodium butyrate (0.5 mM), serum-free medium (SFM), and 5-BrdU (2.5 μM) were added individually 4 days after plating. The drug and medium were changed every 2–3 days and the cultures were maintained for 10–13 days. Contol culture (A) shows that cells grow in clumps and exhibit no spontaneous morphological differentiation (cytoplasmic processes greater than 50 μm in length). Papaverine-treated culture

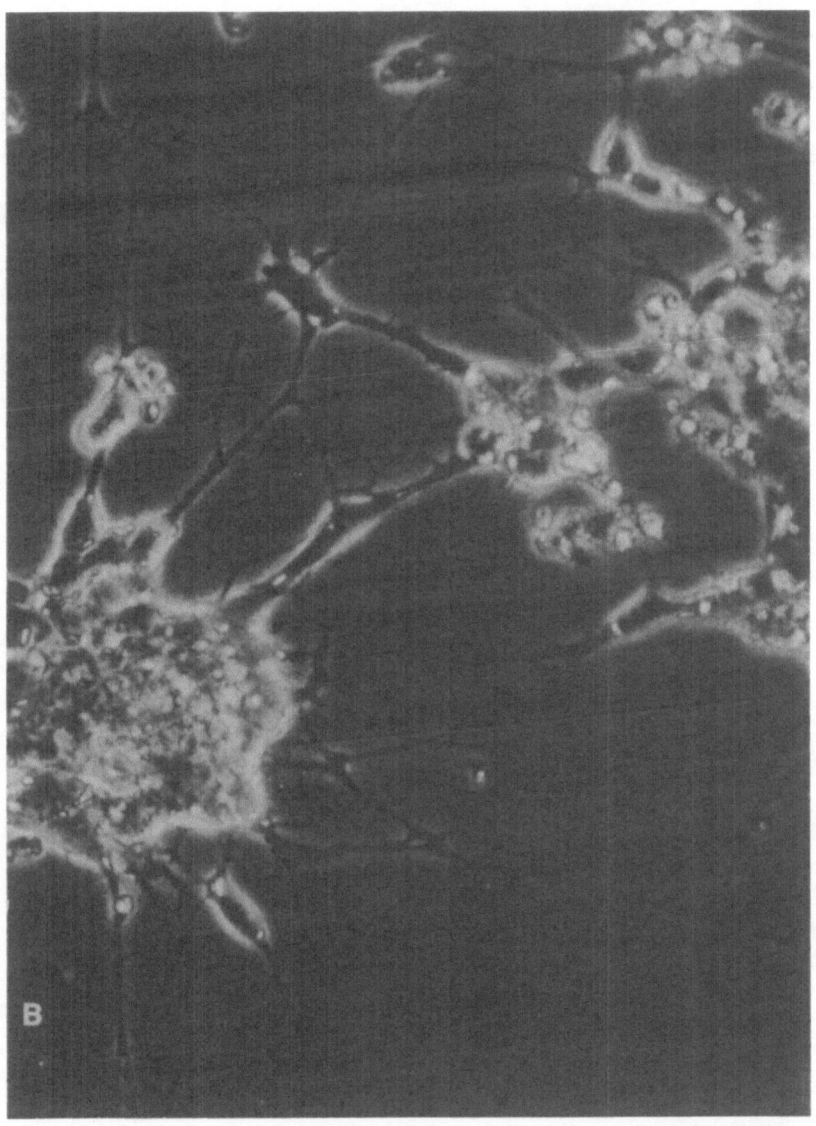

10 days after treatment (B) shows the formation of extensive neurites. Many cell deaths occurred during this period. Sodium butyrate-treated culture 10 days after treatment (C) also shows the formation of extensive neurites. A lot of cell deaths occurred during this period. SFM-treated culture 3 days after treatment (D) and 5-BrdU-treated cultures 10 days after treatment (E) show an extensive neurite formation (From Prasad and Kumar.[19])

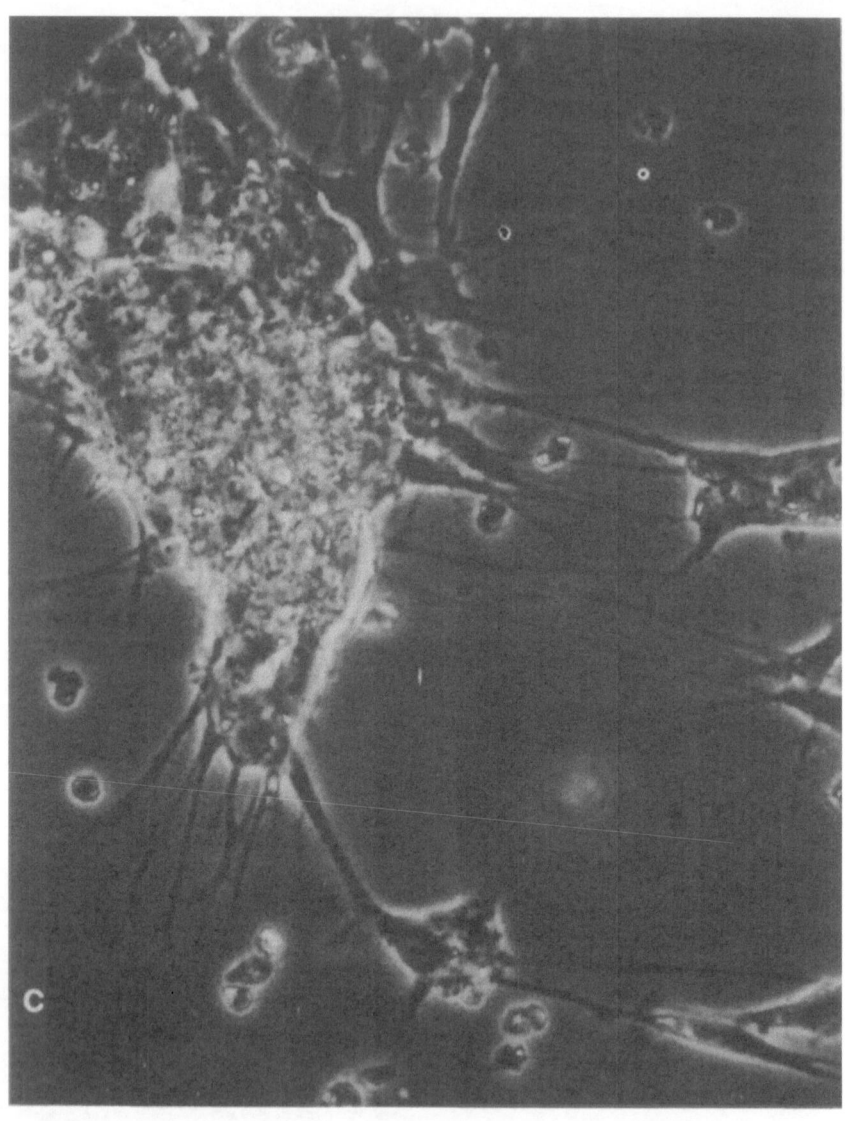

Figure 3-6C

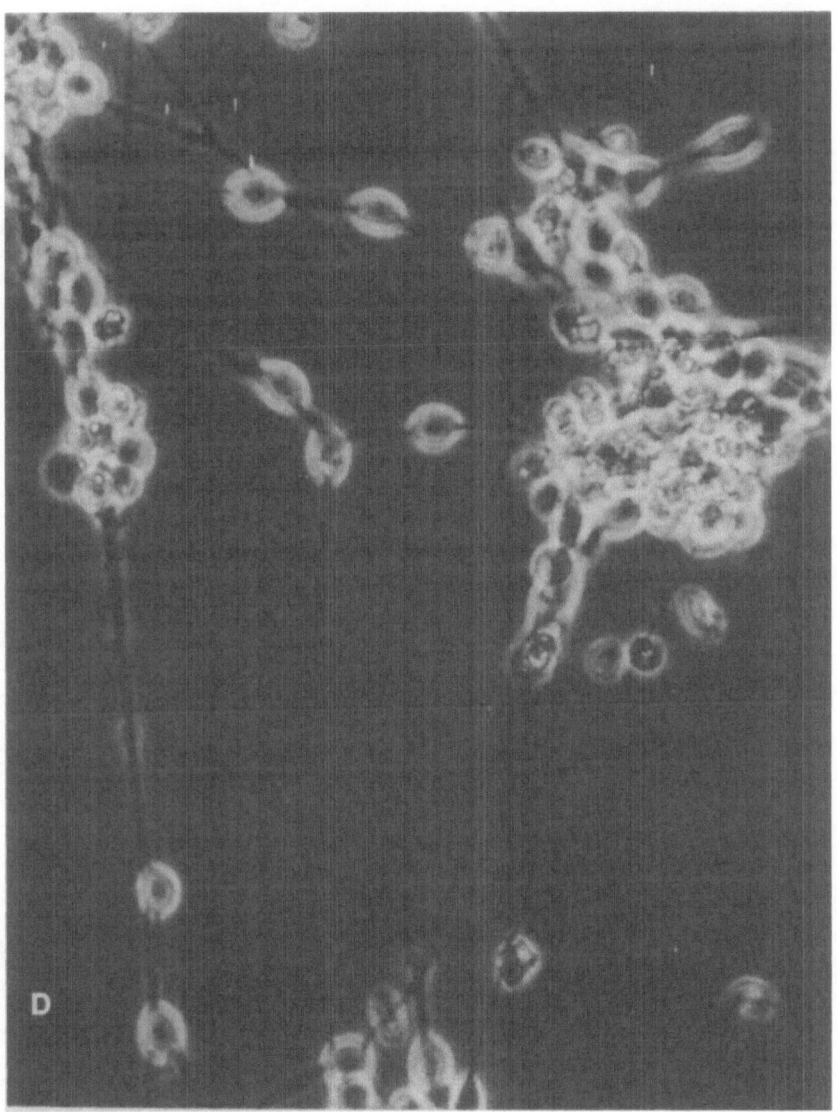

Figure 3-6D

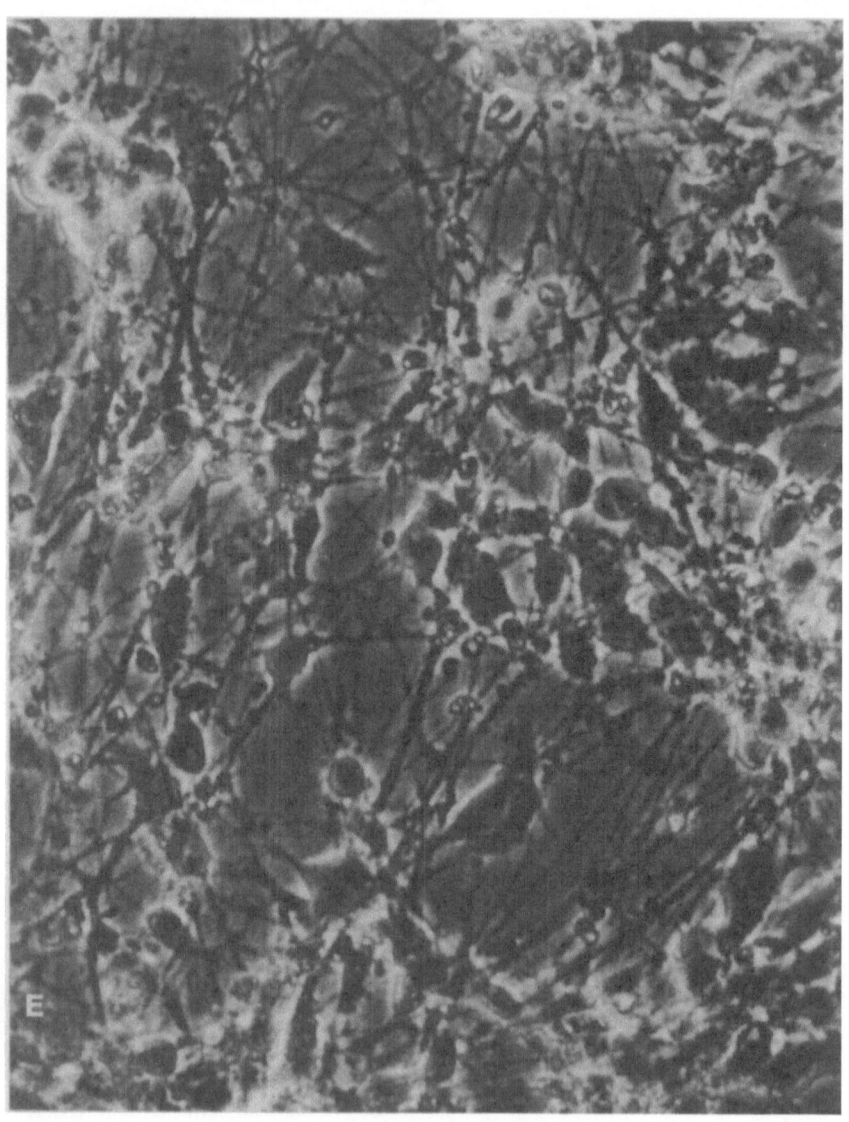

Figure 3-6E

by disrupting 40- to 80-Å diameter microfilaments.[39] Cytochalasin B and vinblastine sulfate block Ro 20-1724-induced neurites in mouse neuroblastoma cells; however, these agents do not diminish the Ro 20-1724-stimulated increase in the cAMP level,[40] indicating further the requirement of the organization of microtubules and microfilaments for the expression of neurites. The expression of this differentiated phenotype does not require the synthesis of new RNA, since cAMP-stimulating agents induce neurites in the presence of actinomycin D (inhibits RNA synthesis)[20] and in enucleated neuroblastoma cells.[41] However, there is controversy regarding the requirement of new protein synthesis. Neurites induced in serum-free medium do not require the synthesis of new protein[24] but 5-bromodeoxyuridine (5-BrdU) and cAMP-induced neurites do.[20,26] This difference may be due to differences in experimental conditions. It has been suggested that neurite formation can be initiated in the absence of protein synthesis but soon reaches a point beyond which *de novo* synthesis of protein is required.[42] Neurite formation in some mouse neuroblastoma cell lines requires strong interaction between the cell and the surface of the culture dish,[26] although this may not be equally true in another mouse neuroblastoma clone.[43] A sustained level of phosphorylated tubulin is not required for the expression of neurite extension.[44]

Neurites, which are formed after treatment of neuroblastoma cells with various agents, may be reversible or irreversible after the removal of the drug. In a sensitive clone, cAMP produces neurites which are irreversible. The irreversibility of neurite formation after treatment with agents that elevate the intracellular level of cAMP is confirmed in some clones by the following observations: When the differentiated cells, 4 days after treatment with prostaglandin $E_1$ ($PGE_1$), a stimulator of adenylate cyclase activity, or Ro 20-1724, are removed from dishes by means of Viokase solution and replated in separate dishes, cells attach and form long neurites within 24 hr even though no drug is present in the medium.[19] The number of morphologically differentiated cells in the newly plated dishes is similar to that in dishes from which the drug is not removed. This indicates that the cellular factors that control the expression of differentiated phenotype remain functional after subculture. It is essential that the culture be maintained in the presence of drugs ($PGE_1$ and Ro 20-1724) for at least 4 days before the irreversibility of neurites can be observed. Analysis of many studies has shown that the difference in the cAMP effects on neuroblastoma observed in various laboratories may be due to the following reasons: (1) clonal difference;

(2) type of drug, and (3) type of serum: Many clones are sensitive to PGE$_1$ and Ro 20-1724 in causing neurite formation. However, some clones, irrespective of their neuronal cell type, are sensitive to PGE$_1$ but not to phosphodiesterase inhibitor and vice versa.[43] The clone insensitive to inhibitors of cAMP phosphodiesterase is also unresponsive to dibutyryl cAMP. Thus, a difference in clone may account for the difference in results. Although we first demonstrated the effect of cAMP on differentiation of neuroblastoma cells in culture by using dibutyryl cAMP, this may not be the best agent to demonstrate the effect of cAMP on neurite formation for the following reasons. (1) The purity of dibutyryl cAMP varies from batch to batch, even though it is obtained from the same commercial source. In addition, the formed butyric acid in the medium may interfere with the action of cAMP on the assembly of microtubules and microfilaments. (2) Many clones do not show any morphological change after treatment with dibutyryl cAMP. Therefore, it is essential to use PGE$_1$ and inhibitors of cAMP phosphodiesterase, in addition to dibutyryl cAMP, before drawing any conclusion about the role of cAMP in regulating growth and differentiation. Unfortunately, in many studies, the conclusion that cAMP is effective, ineffective, or partially effective in producing changes on growth rate and/or differentiation have been based on the use of dibutyryl cAMP alone. The period of treatment of culture with cAMP is very important in causing irreversible neurite formation. It is essential to treat the cells for at least 4 days. The reversibility of neurite formation in another clone after treatment with dibutyryl cAMP may in part be due to the fact that the culture was treated for only 48 hr.[45] The type of serum may be one of the important factors in producing variations in the expression of neurite formation in the same clone.[15]

The inhibition of cell division is not a prerequisite for induction of neurites. On the contrary, neurites are formed in the dividing mouse neuroblastoma cells. cAMP[20] induces neurites prior to the inhibition of cell division. 5-BrdU induces neurites at a concentration which does not inhibit DNA synthesis.[26] Several other differentiated functions, such as neural specific enzymes,[11] synthesis of catecholamines,[46] nervous-system-specific protein,[47] are associated with an increase in the intracellular level of cAMP, and the stimulation of adenylate cyclase activity by neurotransmitters[48,49] can be expressed in dividing nerve cells, though some of these functions may be present at extremely low levels. They may, however, fully express after cessation of cell division, at which time other new differentiated functions may also appear.

cAMP and nerve growth factor (NGF) are two physiological substances which induce neurites in some clones of mouse and human neuroblastoma cells in culture. Dibutyryl cAMP causes neurite formation in some clones of chemically induced central nervous system tumor cells.[50] cAMP and NGF may also be important in the regulation of neurite formation during development of nerve cells. For example, NGF is essential for the growth and differentiation of sympathetic ganglion cells *in vivo*[51] and in culture.[52] Dibutyryl cAMP induces neurite formation in cultures of dorsal root ganglia from chick embryo,[53] mouse sensory root ganglia,[54] and fetal rat brain.[55] Recently it has been shown that dibutyryl cAMP induces neural differentiation in cultures of undetermined presumptive epidermis from three amphibian species.[2] However, the relationship between NGF and cAMP remains poorly understood. A study has shown[55] that NGF (2.5 S) produced a severalfold increase in cAMP level in rat superior cervical ganglia in organ culture within 4 min, which returned to the basal level in about 10 min. The understanding of their interaction may provide a better insight into the regulation of neurite formation and other differentiated functions in neuroblastoma cells. This section has been discussed in detail in Chapter 4.

The expression of neurites appears to be independent of malignancy, since neurites of varying sizes may be formed in the dividing cells. A neuroblastoma cell may permanently stop cell division without the expression of long neurites. The neurites are also expressed in the absence of any increase in the expression of many biochemical differentiated functions.

## Ultrastructural Changes

The ultrastructural features of neuroblastoma ($N_2a$) cells grown in suspension and monolayer were examined.[56] Mouse neuroblastoma cells have several morphological features similar to those of neurons. When grown in monolayer cultures, long arborized neurites are extended that are similar in appearance to the axons and dendrites of normal neurons. These cells also contain organelles and grouping of organelles typically found in nerve cells.

### Filamentous Elements

Filamentous elements are the most prominent structures in the axons and dendrites of normal neurons and generally are oriented parallel to the long axis of these processes. They include microtubules (about 250 Å in diameter), neurofilaments (about 100 Å in diameter), and microfilaments (about 50–60 Å in diameter). These filamentous structures have a similar distribution in neuroblastoma cells, although the neurofilaments may be grouped into large bundles in both the cell bodies and the processes. The neurofilaments observed in neuroblastoma cells have a periodicity of approximately 100 Å and this periodicity may reflect the arrangement of the globular subunit into either a helically coiled thread or into four to six 30-Å-filaments.[56] The finer, 50- to 60-Å microfilaments occur beneath the cell membrane and fill the microspike in the neuroblastoma cells and are probably comparable to those that compose the network in the growth cones and associated microspikes of cultured neurons and to those that occur in a wide variety of cell types. The microfilaments immediately beneath the membrane of neuroblastoma cells can be detected with heavy meromyosin and therefore may be actinlike substance.[57]

*Neuronal Vesicles*

Several types of vesicles found in normal neurons are also found in neuroblastoma cells. One type, the coated vesicle, is found frequently in the perikaryon and neurites of the neuroblastoma cells and appears to be formed by pinocytosis. It has been suggested[58] that the pinocytotic formation of coated vesicle at the nerve terminal of a normal neuron may represent a mechanism for the retrieval of synaptic vesicle membrane from the plasma membrane. Another type of vesicle seen in neuroblastoma cells has a dense core. These vesicles, which are about 1500–2500 Å in diameter, have been previously described in neuroblastoma cells.[42,59] Catecholamine storage was examined in culture of mouse neuroblastoma cell line (N-TD$^6$) using histofluorescence and electron-microscopic, isotopic, and radioautographic criteria.[60] When examined in *para*-formaldehyde-induced histofluorescence, a small percentage of cells in the population show intense catacholamine fluorescence, often localized within discrete regions of the cellular processes. Electron-microscopic examination of these cells reveals both electron-luscent vesicles and, more frequently, electron-dense granules, 50–70 nm and 100–300 nm in diameter, respectively. The distribution of these granules and vesicles varies, but they appear most numerous near the cell surface, along processes and within process endings. By labeling cells with [$^3$H]-dopamine and then allowing the cell to release unbound label in the presence of unlabeled dopamine, the localization of catecholamines stores was visualized by radioautographic technique. While a variety of intracellular distributions of radioactivity was observed, the most prominent concentrations were found in the processes and their terminals; no labeled material was retained when reserpine was present during uptake. Reserpine is known to deplete catecholamines from the storage granules. Thus the topographic coincidence of granules, catecholamine fluorescence, and [³H]dopamine retention in these neuroblastoma cells suggest that catecholamines are stored within these granules in a manner analogous to that observed in normal adrenergic neurons.[60] The neurites of neuroblastoma cells have morphological features generally typical of both axons and dendrites.[56] Dendrites have been characterized as having free ribosomes, smooth reticulum, mitrochondria scattered throughout the cytoplasm, and short cytoplasmic spines emerging from the main dendritic trunk. In contrast, axons have few or no ribosomes beyond the axon billock and initial segment.[61]

Neuroblastoma cells also appear to have membrane specializations, characterized by the presence of dense materials on the cytoplasmic sides of opposed membrane.[56] The function of these membrane specializations is unknown, but they may represent either early stages of synapse formation or nonsynaptic adhesion junctions.

*Electrophysiological Properties of Neurites*

Treatment of cultured mouse neuroblastoma cells with dibutyryl cAMP yields a nondividing cell population of highly electrically excitable cells.[62] Such cells may be a useful model for studying the electrophysiology of single and interacting neurons for biochemical and electrophysiological examination of excitable membrane and for studying regulation of ionophores.

## Synapse Formation

Clonal mouse neuroblastoma × rat glioma hybrid cells (NG 108-15) were shown to form chemical synapses with cultured mouse striated muscle cells.[63] The properties of the synapses between hybrid and muscle cells were similar to those of the normal neuromuscular synapse at an early stage of development. It has been shown[64,65] that at an early stage in the development of synapses between normal mammalian motor neurons and striated muscle cells, a single muscle cell is innervated by more than one neuron, whereas at a later developmental stage only one neuron synapses with a single muscle cell. The efficiency of synaptic communication during the early stage in synapse formation is low, and most muscle cell responses are below the threshold for activation of action potentials.[64-66] At a later stage in the development of the synapse the efficiency of transsynaptic communication becomes 100%, since every neuronal action potential elicits muscle cell contraction. The number of synapses formed and the efficiency of transmission across synapses in neuroblastoma × glioma hybrid cells were found to be regulated, apparently independently, by a component in the culture medium. Under appropriate conditions synapses were found with 20% of the hybrid-muscle cell pairs examined, indicating that the hybrid cells form synapses with relatively high frequency. The frequency of synapse formation between hybrid and muscle cells equals or exceeds the synapse frequency reported for normal dissociated spinal-cord neurons and striated muscle cells *in vitro*.[67] The efficiency of synaptic transmission might be affected by many factors. For example, high concentrations of nicotinic acetylcholine receptors have been found at points of contact between neuroblastoma and striated muscle cells.[68] However, similar receptor "hot spots" are found on muscle cells in the absence of neuroblastoma cells.[69,70] Establishment of synapses between nerve and muscle poses a paradox. On the one hand, muscle movements are highly coordinated, which suggests that neuromuscular synapses and other synapses in the neural circuits are assembled with high precision; on the other, the demonstrated ability of autonomic neurons of the vagus,[71] sympathetic ganglion neurons,[72] and neuroblastoma hybrids[63] to synapse with striated muscle suggests that functional synapses can form that may not be dependent upon highly specific cell recognition molecules. Although most striated muscle cells in mammals are innervated by spinal motor neurons, it is interesting to note that the esophagus contains striated muscle cells normally innervated by autonomic neurons of the vagus. The molecular nature of the cell interactions that lead to synapse formation is unknown. It is clear that neuroblastoma × glioma hybrids adhere firmly to muscle cells[63] as well as to one another, to fibroblasts, to other cell types, and to the polystyrene petri dish. Whether the hybrid cells have specific cell-recognition molecules required for synapse formation remains to be determined.[63] It has been postulated[63] that the hybrid cells may constitute one class of cells with respect to synapse formation, and muscle cells another class; and any well-differentiated hybrid cells may be able to form synapses with any muscle cell in culture. If specific cell-recognition molecules are required to establish synapses between hybrid cells and mouse striated muscle cells, it seems likely that few kinds of recognition molecules may be required to establish a connection during the early stage of development.[63]

## Membrane Changes

Mouse neuroblastoma cells show strong agglutination in the presence of concanavalin A (Con A) or wheat-germ agglutinin (WGA), whereas differentiated cells, induced by inhibitor of cAMP phosphodiesterase (Ro 20-1724), show very little.[73] However, differentiated cells, induced by dibutyryl cAMP or PGE₁, show agglutination in the presence of Con A and WGA to an extent similar to that observed in malignant neuroblastoma cells. Thus, changes in the agglutination sites are not necessarily linked with the differentiation of neuroblastoma cells. Therefore, it is unlikely that reverse changes can be linked with malignant transformation of nerve cells. Glycopeptides obtained from the surface membrane of neuroblastoma derived from those clones which did not differentiate resembled the glycopeptides obtained from the surface of hamster embryo cells transformed by virus.[74] Conversely, the glycopeptides obtained from either of the clones that extended neurites resembled the glycopeptides from nontransformed cells. It has been reported[75] that the polypeptides of the surface membrane fraction of both control and dibutyryl cAMP-induced morphologically differentiated mouse neuroblastoma cells are similar. However, a protein of molecular weight 78,000 was demonstrable on the surface of differentiated cells and was not demonstrable either in malignant cells or in suspension cultures of cells treated with dibutyryl cAMP. This shows that the expression of the morphological phenotype is linked with the appearance of new polypeptides on the surface membrane.

A cell-surface glycopeptide characteristic of differentiated neuroblastoma cells (clone I) has been identified. 5-BrdU increased the synthesis of this glycopeptide. This effect of BrdU is reversed by the simultaneous presence of thymidine, indicating the BrdU effect is on the transcriptional level.

## POLYUNSATURATED FATTY ACID METABOLISM DURING DIFFERENTIATION

Neuroblastoma cell culture (N18) takes up linoleic and linolenic acids and esterifies them at approximately the same rate.[77] Between 80 and 95% of total radioactive counts, representing 50–70% of the administered dose, are incorporated into cellular phospholipids after a 48-hr incubation. The distribution of radioactive label between the

phospholipid fraction differs for the two essential fatty acids employed as precursor. When linoleic acid was the precursor, choline
phosphoglycerides (CPG) was the major labeled lipid, while with
linolenic acid as precursor the ethanolamine phosphoglycerides
(EPG) fraction was labeled most extensively. The marked difference
between the labeling pattern of CPG and EPG suggests that a direct
transfer of $^{14}$C-labeled fatty acids from CPG to EPG is not a major
path in neuroblastoma cells. When cell division ceases after treatment
with cytosine arabinoside, the pattern of fatty acid incorporation
changes. The diminished entry of label into EPG and CPG in the
presence of cytosine arabinoside suggests that the turnover of these
lipids is reduced when the cell division stops. Since extensive neurites are formed after cytosine arabinoside treatment,[30] this finding
suggests that these lipids are involved in the formation of new cell
membranes. But this is contradicted by the data on cells treated with
dibutyryl cAMP and serum-free medium. Treatment of cells with
dibutyryl cAMP or by withdrawal of serum fails to alter the extent of
fatty acid incorporation and its distribution among various lipid fractions. Within the EGP fraction, phosphatidyl ethanolamine (plas-
malogen) is the major acceptor for the higher polyunsaturated fatty
acids derived from linolenic acid. When cell division is slowed down
or arrested by removal of serum or addition of cytosine arabinoside,
the relative amount of labeled polyunsaturated fatty acids in plasmalogens is increased. This suggests[77] that plasmalogens may be in-
timately involved in cellular differentiation and in the formation of
new cell processes.

## MODIFICATION OF GENE EXPRESSION BY cAMP

An elevation of cAMP in certain clones of neuroblastoma cells in
culture by $PGE_1$ or by inhibitors of cyclic nucleotide phosphodiesterase or by analogs of cAMP causes an extensive modification of
gene expression that appears to be linked with the degree of differentiation. The expression of certain genes is increased, whereas the
expression of others is decreased.[1,78,79] The regulation of gene expression has been studied by measuring the individual gene product.

### Neurotransmitter Metabolizing Enzymes

Because of the importance of neurotransmitters in neuronal function, the regulation of neurotransmitter metabolizing enzymes has
been extensively studied using a variety of experimental models.

Naturally, when the clonal lines of nerve cells in culture became available, the regulation of these enzymes was extensively studied using mouse neuroblastoma, human neuroblastoma, and rat CNS tumor cells.

## Tyrosine Hydroxylase

Tyrosine hydroxylase (TH) is the rate-limiting enzyme in the biosynthesis of catecholamines. The activity of TH has been demonstrated in various neuroblastoma clones.[3,11,19,50,80] TH activity in some clones is similar to that found in brain. The role of cAMP in the regulation of TH activity has been studied using three different model systems of neuroblastoma cells in culture, normal nervous tissue, and synaptosomal preparation. Although cAMP-stimulating agents increased TH activity in these systems, other agents which do not increase cAMP level also increase TH activity. Dibutyryl cAMP increases TH activity in concentration and tissue-dependent fashion (Figure 3-7). A significant increase in the enzyme activity was observed at 3 days after treatment.[12] The enzyme activity did not significantly decrease when the drug was removed at 3 days after treatment, and the enzyme activity was assayed 2 days later. This indicated that dibutyryl cAMP-induced elevation of TH activity is irreversible, at least during a period of 2 days. Sodium butyrate, a

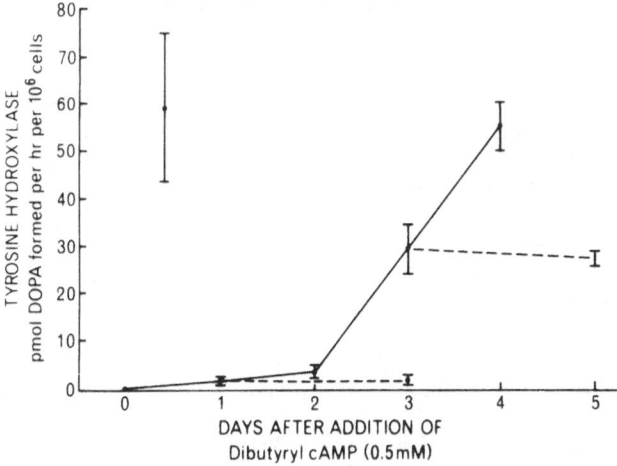

**Figure 3-7.** TH activity in neuroblastoma tumor freshly removed from mice (top, left) and in mouse neuroblastoma cells in culture for at least 100 days after exposure to 0.5 mM $N^6,O^2$-dibutyryl cAMP for various period of time (solid line). In some studies, after 1 or 3 days the medium was replaced by fresh medium lacking the cyclic nucleotide, and the incubation was continued for two days (dashed line). (From Waymire et al.[12])

degradative product of dibutyryl cAMP in solution also increased TH activity but to a lesser degree (Table 3-3). However, sodium butyrate increased the intracellular level of cAMP by about twofold.[81] Other analogs of cAMP, such as $N^6$-monobutyryl-3',5'-cAMP, 8-amino methyl 3',5'-cAMP, and 8'-methylthio-cAMP at a concentration of 0.3 mM, markedly elevated TH activity. Papverine, an inhibitor of cyclic nucleotide phosphodiesterase, also markedly increased TH activity. All the above agents inhibit the growth rate; therefore, it is not clear if TH activity is regulated by cAMP or by growth rate. X irradiation, 6-thioguanine (6-TG), and confluency, which inhibit the growth rate without changing the intracellular level of cAMP,[81] do not increase TH activity; 5-BrdU, which increases the cAMP level by about two-fold,[81] increases TH activity by about fourfold[29] (Table 3-4). 5'-AMP and 3',5'-cAMP, which inhibit cell division, do not increase TH activity.[12] These data suggest that the inhibition of growth rate is not sufficient for increasing the TH activity. However, Amano et al.,[3] reported that TH activity increases as the rate of cell division decreases. This difference may be due to a difference in cell lines. A twofold increase in TH activity of rat CNS tumor cells occurs when they reach the stationary phase.[50] The effect of serum-free medium on TH activity is complex and controversial. Serum-free medium increases TH activity in one clone of neuroblastoma cells,[24] but it does not do so in other clones.[30,82] Although serum-free medium increases the intracellular level of cAMP by about twofold,[81] it fails to increase

**Table 3-3.** TH Activity and Protein Concentrations in Mouse Neuroblastoma Clone NBP$_2$ after Various Treatments[a]

| Condition[b] | TH[c] | Protein[d] |
|---|---|---|
| Control, log phase[e] | 15.1 ± 1.9 | 1.6 ± 0.2 |
| Control, confluent phase[f] | 11.2 ± 0.7 | — |
| Serum-free medium | 17.3 ± 0.4 | — |
| Dibutyryl cAMP, 0.5 mM | 473 ± 17 | 3.11 ± 0.33 |
| 8-Methylthio-cAMP, 0.3 mM | 587 ± 9 | 2.95 ± 0.28 |
| Papaverine, 0.13 mM | 977 ± 46 | 3.00 ± 0.23 |
| Sodium butyrate, 0.5 mM | 300 ± 12 | 2.80 ± 0.18 |

[a] From Waymire et al.[12]
[b] All drug treatments are started 1 day after replating and continued for 3 days.
[c] Picomoles of product formed in 30 min/10$^6$ cells.
[d] Milligrams per 10$^6$ cells.
[e] Three days after replating.
[f] Grown in the absence of newborn calf serum for 3 days.

**Table 3-4.** Effect of Various Analogs of Nucleic Acid Bases on the Levels of Neural Enzymes in Neuroblastoma Cells[a]

| | | Enzyme activity (pmol/15 min per $10^6$ cells) | | |
|---|---|---|---|---|
| Clone | Enzyme | Control | G-TG | 5-BrdU |
| $NBA_{2(1)}$ | COMT | 26 ± 2.3[b] | 64 ± 3 | 40 ± 1.6 |
| $NBA_{2(1)}$ | TH | 1.3 ± 0.7 | 1.6 ± 0.5 | 5.8 ± 0.89 |
| $NBE^-$ | CAT | 190 ± 39 | 696 ± 102 | 347 ± 80 |

[a] Cells (0.5 × $10^6$) were plated in 75-cm² Falcon plastic flasks, and 6-thioguanine (6-TG, 0.5 $\mu$M) and 5-BrdU (5 $\mu$M) were added separately 24 hr later. The fresh drug solution and growth medium were added 2 days after treatment and catechol-$O$-methyltransferase (COMT) and TH and CAT activities were assayed 3 days after treatment according to the method of Axelrod et al. (1958), Waymire et al. (1971), and Fellman (1969), respectively. Each value represents an average of six to eight samples.
[b] S.D.

TH activity in the same clone. This indicates that the presence of serum is necessary for the effect of cAMP on TH activity.

Dibutyryl cAMP and sodium butyrate also increased TH activity in murine neuroblastoma[83] and in human neuroblastoma cells of IMR-32 line (Table 3-5). In transplantable mouse testicular teratoma (OTT 6050) an increase in the level of cAMP correlated with an enhanced activity of TH with increased proportion of neuroepithelial cells.[84] The mechanism of cAMP effect on TH activity is unknown. Cycloheximide, an inhibitor of protein synthesis, inhibits completely the dibutyryl cAMP-induced increase in TH activity, indicating

**Table 3-5.** Effect of Various Agents on the TH Activity in Human Neuroblastoma Cell Culture[a]

| Treatment | TH activity (pmol/30 min/mg protein) |
|---|---|
| Control | 5 ± 2.0[b] |
| Dibutyryl cAMP | 550 ± 102 |
| Sodium butyrate | 72 ± 19 |

[a] Cells were plated in 75-cm² Falcon plastic flasks, and dibutyryl cAMP (0.5 mM) and sodium butyrate (0.5 mM) were added separately 5 days later. Fresh medium and drug were replaced 2–3 days after treatment. The TH activity was measured 10 days after treatment. Each value represents an average of eight to ten samples. From Prasad and Kumar.[19]
[b] S.D.

perhaps that dibutyryl cAMP affects TH activity at the translational level. It has been shown that dibutyryl cAMP increases tyrosine uptake in neuroblastoma cells.[85,86] The increase in tyrosine hydroxylation by dibutyryl cAMP-treated cells was in direct proportion to the increase in tyrosine uptake. The rate of tyrosine hydroxylation per molecule of substrate did not change by dibutyryl cAMP treatment, thus indicating that tyrosine hydroxylation increased primarily due to the increase in tyrosine uptake. At this time, it is unknown if the classical mechanism of cAMP effect, which involves phosphorylation of protein by cAMP-dependent protein kinase, is operative for this particular modification of gene expression.

In contrast to mouse and human neuroblastoma cells, dibutyryl cAMP did not increase TH activity of rat CNS tumor cells in culture,[50] although it did cause elongation in neurite outgrowth. This discrepancy cannot be explained at present. However, the effect of dibutyryl cAMP alone may not be indicative of cAMP-effect. Therefore, the role of cAMP in the regulation of TH activity in rat CNS tumor cells remains inconclusive until the effect of at least one inhibitor of cyclic nucleotide phosphodiesterase and one stimulator of adenylate cyclase of TH activity and on cAMP-dependent phosphorylation activity is investigated. Recent studies indicate that 5-(3,3-dimethyl-1-triazeno)imidazole-4-carboxamide (DTIC) increases TH activity by sevenfold 3 days after treatment; however, it does not change the level of cAMP.[87] Unlike in mouse neuroblastoma cells, 5-BrdU increases TH activity in human neuroblastoma cells in culture without changing the intracellular level of cAMP.[27] These data suggest that the regulation of TH activity in neuroblastoma cells may involve at least two modes one of which involves the changes in the cAMP level.

Thus, DTIC and 5-BrdU increase TH activity without changing the level of cAMP. Therefore, the suggestion that the regulation of TH involves more than one mode remains speculative until it is shown that these agents do not increase cAMP-dependent phosphorylation activity. Dibutyryl cAMP increases the TH activity in cultures of sympathetic ganglia[88] and in striatal slices obtained from rat.[89] Dibutyryl cAMP elicits a concentration-dependent stimulation of TH activity in the striatal and mesolimbic synaptosomes.[90] The percent stimulation in enzyme activity is significantly higher in the mesolimbic synaptosomes than in the striatal synaptosomes. Dibutyryl cAMP and depolarizing agents (ouabain or veratridine) produce an additive effect on synaptosomal TH activity, indicating that they stimulate TH activity by different mechanisms. cAMP does not stimulate soluble striatal TH activity unless it is added in combination with ATP and $Mg^+$

compounds required for the activity of cAMP-dependent protein kinase. The cAMP-induced increase in soluble TH activity depends upon the concentration of added protein kinase and upon the pH of the reaction mixture. Dibutyryl cAMP has the same effect on the kinetic of TH in synaptosome as cAMP on the soluble TH. The nucleotide does not alter the apparent $K_m$ for tyrosine, reduces the $K_m$ for petridine cofactor, and increases the $K_i$ for dopamine. Thus cAMP increases the affinity of TH for the petridine cofactor and concomitantly decreases the affinity for the end-product inhibition.

Most of the data indicate that cAMP may be one of the important factors in the regulation of TH activity during differentiation; however, more data are needed before any conclusion can be made.

*Dopamine β-Hydroxylase*

The last step in the biosynthesis of norepinephrine (NE), the hydroxylation of dopamine (DA), is catalyzed by the enzyme dopamine β-hydroxylase (DBH). DBH activity is demonstrable in both mouse[91] and human[4] neuroblastoma cells in culture. Dibutyryl cAMP increases DBH activity in only one of two hybrid clones (mouse neuroblastoma × rat glioma) which contain DBH activity, although it causes neurite formation in both clones.[92] This means that in the hybrid cells the extension of neurites is not necessarily coupled with the increase in DBH activity. In mouse teratoma, the increased level of cAMP correlates well with the increased activity of DBH.[84] When rat superior cervical ganglia were incubated with dibutyryl cAMP, DBH activity increased approximately twofold over a period of 6 hr.[93] A study[94] shows that dibutyryl cAMP had two effects on the NE-synthesizing system of isolated rat superior cervical ganglia. It raised ganglionic NE levels by increasing the synthesis of NE from tyrosine but not from dihydroxyphenylalanine (dopa). It also increased DBH activity. The effect of dibutyryl cAMP on DBH activity was blocked by cycloheximide, but the effect on NE level was unaffected. In addition, dexamethasone also increased DBH activity without altering NE level. These data suggest that the regulation of the level of DBH activity and NE in ganglion cells are apparently distinct. The increase in DBH activity caused by dibutyryl cAMP, like the increase during increased nervous activity *in vivo*, is blocked by cycloheximide, suggesting that it involves the synthesis of new enzyme proteins. This also shows that dibutryryl cAMP probably affects DBH activity at the translational level. It is still premature, then, to suggest the role of cAMP in the regulation of DBH activity until the criteria mentioned in the Introduction section are met.

In the peripheral sympathetic nervous system an increased activity of the preganglionic cholinergic fibers leads to an increased synthesis of tyrosine hydroxylase and dopamine $\beta$-hydroxylase in the adrenergic neurons and adrenal chromaffin cells.[95,96] In sympathetic ganglia this induction can be abolished by transecting the preganglionic cholinergic fibers[97,98] or by administering nicotinic blocking agents.[99] Moreover, high doses of ACh or carbachol have been reported to lead to an increase in tyrosine hydroxylase activity in denervated adrenals.[100-101] These findings suggest that ACh acts as the first messenger in transsynaptic induction of tyrosine hydroxylase and dopamine $\beta$-hydroxylase. The fact that stimulation of preganglionic cholinergic fibers produces a marked increase in cAMP in the rabbit superior cervical ganglia,[102,103] and that high concentrations (1–5 mM) of dibutyryl cAMP produce a cycloheximide-sensitive increase in tyrosine hydroxylase in mouse superior cervical ganglia in organ culture[88] could be taken as an indication that this nucleotide acts as a second messenger in transsynaptic enzyme induction. It has been reported[101,104] that an experimental condition became apparent in which there is a correlation between the rate of increase in cAMP and subsequent tyrosine hydroxylase induction in the rat adrenal medulla. However, the rise in cAMP in rabbit superior cervical ganglion after preganglionic stimulation appears to be mediated by a sequential muscarinic-dopaminergic mechanism[102,103] whereas transsynaptic induction is mediated by a nicotinic mechanism.[99] Moreover, concentrations of dibutyryl cAMP leading to increases in enzyme activities in sympathetic ganglia in organ culture do not discriminate between those enzymes specifically increased by transsynaptic induction *in vivo*, tyrosine hydroxylase and dopamine $\beta$-hydroxylase, and enzymes not increased under these conditions.[105] To decide whether cAMP plays a role as second messenger in the transsynaptic induction of tyrosine hydroxylase, it is desirable to discriminate between neuronal and extraneuronal changes in cAMP concentration. Treatment of newborn rats with NGF antiserum or 6-hydroxydopamine (6-OHDA), leading to destruction of 61–85% of the adrenergic nerve cell bodies in the superior cervical ganglion, led to a decrease in cAMP of only 16–28%.[106] This observation demonstrates that a relatively small portion of cAMP is localized in the adrenergic neurons. However, administration of isoproterenol produced a 12-fold increase in cAMP only in this neuronal pool. Neither single nor repeated injections of isoproterenol led to the induction of tyrosine hydroxylase. These data suggest that cAMP may not act as a second messenger in the transsynaptic induction of tyrosine hydroxylase in the rat superior cervical ganglion. In the rat medulla, treatment with reserpine led to both a short-lasting (60–90 min) increase in cAMP and a subsequent induction of tyrosine hydroxylase.[106] However, the increase in cAMP was almost completely prevented (40% compared to 32%) by pretreatment of the rats with propranolol, while the induction of tyrosine hydroxylase was not inhibited. This observation also argues against an exclusive key function of cAMP in transsynaptic induction of tyrosine hydroxylase in the adrenal medulla.

### Choline Acetyltransferase and Acetylcholinesterase

Choline acetyltransferase (CAT) synthesizes acetylcholine (ACh), and acetylcholinesterase (AChE) degrades it. The CAT activity in neuroblastoma cells increases after treatment of cells with cAMP-stimulating agents.[107] However, the enzyme activity is increased also after treatment of cells with agents which do not increase cAMP level (Table 3-6). All these agents inhibit cell division; therefore, it was suggested[107] that the activity of CAT is inversely related to growth rate. 5-BrdU and 6-TG increase CAT activity in mouse NBE⁻ clone by about two- to fourfold.[29] 5-BrdU on CAT activity in mouse neuroblastoma cells is different from that in human cells. The reason for this

**Table 3-6.**  Effect of Various Agents on CAT Level of Neuroblastoma Cells[a]

| Treatment | Activity (pmol/15 min per $10^6$ cells) |
|---|---|
| Control (exponential) | 260 ± 35[b] |
| Control (confluent) | 300 ± 34 |
| Dibutyryl cAMP (0.5 mM) | 1300 ± 72 |
| PGE₁ (10 μg/ml) | 880 ± 100 |
| Ro 20-1724 (200 μg/ml) | 1280 ± 160 |
| 5'-AMP (0.25 mM) | 1320 ± 80 |
| Butyric acid (0.5 mM) | 760 ± 100 |
| 600 rads | 1640 ± 144 |

[a] Neuroblastoma cells (0.5 × 10⁶) were plated in 75-cm² Falcon plastic flasks, and each drug was added 24 hr later. Fresh growth medium and drug were added 2 days after drug treatment and the CAT was analyzed 3 days later. Each value represents an average of five to six samples. From Prasad and Mandal.[95]

difference in regulation is unknown. Even within mouse clones, there appears to be some difference in the regulation of CAT activity. For example, CAT activity increases when the cells reach the stationary phase of growth.[109] In the clone NBE⁻ (which contains CAT but no TH), however, no such effect was observed. This may be due to a difference in the clone or in the experimental conditions. The CAT activity in dibutyryl cAMP-treated cells markedly increases[107]; however, it has been reported[108] that the enzyme activity decreases under similar experimental conditions. This discrepancy may be due in part to a difference in clone and/or in the presentation of data. We have expressed the enzyme activity as pmol/$10^6$ cells, whereas Simantov *et al.*[109] expressed it as pmol/mg protein. Since the protein contents in dibutyryl cAMP-treated cells increase by three- to fourfold, the CAT activity when expressed as pmol/mg protein may show some decrease. An increase in total cellular protein does not always mean an increase in the specific activity of enzyme. Catechol-*O*-methyltransferase (COMT) activity increases in X-irradiated and sodium-butyrate-treated cells[110] but shows no change in cAMP-induced differentiated cells, although the increase in total protein content is similar in all the treated cells.

Although cAMP-induced differentiated neuroblastoma cells have increased activity of CAT, agents such as X irradiation, 6-thioguanine, and 5'-AMP, which do not increase cAMP level, increase CAT activity. The latter observation suggests that cAMP may not be the only agent involved in the regulation of CAT activity. This cannot be ascertained fully until cAMP-dependent phosphorylation activity after treatment with noncyclic AMP agents does not change. However, in a general sense, the activity of CAT is inversely related to growth of neuroblastoma cells in culture. Whether a similar regulatory mechanism is operative during differentiation of mammalian nerve cells *in vivo* cannot be ascertained.

AChE activity is demonstrable in all types of mouse neuroblastoma clones irrespective of their neurotransmitter synthesizing enzymes.[3,80] Dibutyryl cAMP increases AChE activity in mouse neuroblastoma cells.[21,108,111] Agents which inhibit cell division but do not increase the intracellular level of cAMP also increase AChE activity; therefore, it has been suggested[111,114] that the activity of AChE is inversely related to growth rate. The increase in enzyme activity can be expressed with or without the expression of neurite formation. Serum-free medium increases AChE activity.[30,115] The The specific activity of AChE begins to increase rapidly after an initial lag period of about 2–3 days, reaching a maximum level (10- to 20-fold increase) by 7 days after induction. Cordycepin inhibits mRNA synthesis without affecting the stability of mRNA synthesized before treatment.[116] Cordycepin effectively blocked the increase in the rate of AChE synthesis that occurs as a result of serum deprivation, indicating that the induction process itself requires the synthesis of new mRNA. Rates of reappearance of AChE in neuroblastoma cultures were measured after inhibition of preexisting enzyme by pinacolylmethylphosphonofluoridate (soman). Treatment with 1 $\mu$M soman for 1 hr reduced AChE activity to 3–5% of its initial level in both log-phase and stationary-phase cultures. In order for the rate of appearance of AChE activity after inhibition by soman to be considered a true measure of the rate of AChE synthesis, it was necessary to show that protein synthesis was required and that synthesis of AChE was not accelerated by treatment with soman. Cycloheximide completely inhibited recovery, so neither reactivation of inhibited enzyme nor assembly of active enzyme from a pool of inactive precursor was significant. Return of enzymatic activity after irreversible inhibition of AChE by soman in serum-free-medium-induced differentiated cells was blocked by cycloheximide, but not by cordycepin, suggesting that protein synthesis but not mRNA synthesis was required to replace the enzyme. It has been shown[117] that AChE activity and the number of ACh receptors increased 8- to 10-fold when the cells were induced to differentiate by dibutyryl cAMP. In the change from logarithmic to a stationary phase of growth, there was a 10-fold increase in activity of the enzyme and a 5-fold increase in the number of receptors. There were changes in cell size during growth and during differentiation. When corrections were made for changes in cell size, the increase in the number of ACh receptors per square micrometer of cell surface was 7-fold during the change from a linear to a stationary phase of growth and 4-fold per square micrometer when the cells differentiated. Using antibodies against AChE, a neuroblastoma clone with an 80-fold decrease in AChE, activity was selected. This selection did not result in a decrease in the average number of $3.4 \times 10^7$ ACh receptors per cell. This result suggests that there are different determinants for acetylcholine receptors and AChE. Proteolytic enzymes and compounds that block sulfhydryl groups removed or inactivated 80% of the ACh receptors without decreasing the activity of AChE on the cell surface. This further shows the experimental separation between receptors and the enzyme in neuroblastoma cells. Based on these data it has been suggested[117] that there are different genes for the acetylcholine receptor and AChE and that both are regulated during growth and differentiation by a common regulator gene.

A further study shows[118] that 5-BrdU and dibutyryl cAMP induce AChE activity in neuroblastoma cell by different mechanisms. For example, actinomycin D inhibited 5-BrdU-induced increase in enzyme activity but did not inhibit dibutyryl cAMP-induced increase in enzyme activity. This suggests that dibutyryl cAMP affects AChE activity at the translational level, whereas 5-BrdU affects at the transcription level. The effect to 5-BrdU on AChE is similar to that produced by serum-free medium[115] in which the induction of AChE activity requires the synthesis of new mRNA. The different mechanisms of induction of AChE were further demonstrated by the following observations: (1) Enzyme activity in dibutyryl cAMP-resistant neuroblastoma cells was included by 5-BrdU and not by dibutyryl cAMP; and (2) in the temperature-resistant neuroblastoma cells, induction of 5-BrdU was lower at 40° than at 37°, and induction by dibutyryl cAMP was higher at 40° than at 37°. The difference in induction of AChE activity at two temperatures was associated with a higher inhibition of cell division at 40° by both compounds.

The increased activity of AChE observed in neuroblastoma cells grown *in vitro* is also maintained when they are grown *in vivo*[119]; however, the morphological features are markedly changed; indicating that the morphological behavior *in vitro* cannot be correlated directly with the malignancy *in vivo*.

Although most of the studies indicate that the activity of AChE is inversely related to growth rate, one study[120] indicates that the AChE activity is increased in dividing neuroblastoma cells by acetylcholine. This discrepancy is not yet resolved. The direct role of cAMP in the regulation AChE activity in neuroblastoma cells cannot be ascertained, since many agents that inhibit cell division without changing the cAMP level also increase AChE activity. The cAMP-dependent phosphorylation activity as a function of time after treatment of cells with noncyclic AMP agents must be measured before any conclusion can be reached. If the increase in cAMP-dependent phosphorylation occurs prior to increase in enzyme activity, a possible role for cAMP in the regulation of enzyme activity can be considered. However, it is also possible that the regulation of CAT and AChE involves more than one mode, one of which is alteration in the level of cAMP.

*Catechol-O-methyltransferase*

Human neuroblastoma cells contain a high activity of catechol-*O*-methyltransferase (COMT). The lack of hypertension in patients with neuroblastomas is attributed to the presence of COMT, which degrades norepinephrine before it is released in the circulation.[121] COMT activity has been demonstrated in several lines of mouse neuroblastoma cells[110,122] but is undetectable in human neuroblastoma cell line IMR-32.[27] The COMT activity neither increases in confluent cells[110,122] nor in cAMP-induced differentiated mouse neuroblastoma cells.[110] However, X irradiation (600 rads) and sodium butyrate (0.5 mM) markedly increase the enzyme activity.[110] The increase in COMT activity in the irradiated cells is blocked by cycloheximide and actinomycin D, indicating that X irradiation increases the enzyme activity through changes at the transcriptional level. 6-TG and 5-BrdU also increase COMT activity in mouse cell.[29] These data suggest that COMT activity level may not be strictly linked either with the growth rate or the morphological differentiation. Although COMT activity is undetectable in human neuroblastoma cells (IMR-32 line), it becomes detectable after treatment of cells with 5-BrdU.[27] With the exception of sodium butyrate (mouse and human cells) and 5-BrdU (mouse cells only), all cAMP-stimulating agents fail to increase COMT activity. The mechanism of regulation of COMT activity during differentiation of nerve cells remains unknown.

### Glutamate Decarboxylase

Glutamate decarboxylase (GD) catalyzes the synthesis of γ-aminobutyric acid (GABA). The activity of GD was increased two- to threefold in both NS-20 and NIE-115 clones after the cell was treated with 1 mM dibutyryl cAMP. No change in enzyme activity occurred when the cells were cultured in serum-free medium for about 24 hr.[123,124] Neutral detergents, such as Triton X-100 or Triton CF-54, did not stimulate the membrane-bound enzyme. A cationic detergent, G-3634-A, stimulated the enzyme activity when the cholinergic NS-20 cell homogenate was used as the enzyme source, and inhibited it (20%) in the adrenergic clone. However, anionic detergents such as SDS and taurocholate had little or no effect on GD activity when tested in NIE-115 and NS-20 cells. The treatment of fetal rat brain culture with 1 mM $N^6$-monobutyryl cAMP increased the specific activity of GD by about 2.5-fold, but the enzyme activity in rat glioma cells under a similar experimental condition did not change.[124] On the other hand, in CNS tumor cell line (B65 clone) dibutyryl cAMP induced in 1 day the glutamate decarboxylase I activity before there was any effect of the cyclic nucleotide on cell division.[50] The increase in GD activity correlates with the dibutyryl cAMP-induced morphological changes. In B103 clone dibutyryl cAMP inhibited the switch from the glutamic decarboxylase GD II type to the GD I type of activity normally observed at the cessation of exponential growth.[50]

*Monoamine Oxydase*

Dibutyryl cAMP and depolarizing agent (45 mM K$^+$) increase monoamine oxidase (MAO) activity in explant of chick embryo ganglia.[130] Depolarizing agents are known to increase cAMP level in brain slices.[131] Thus the regulation of MAO by cAMP remains a possibility.

*Thymidylate Synthetase*

Thymidylate synthetase converts dUMP to dTMP. The specific activity of this enzyme increases by about 2.4-fold when confluent cultures are split and begin to divide rapidly.[109] Dividing cells, which are replicating their DNA content, have a greater need for thymidylate, and thus, increased thymidylate synthetase-specific activity could maintain the thymidylate pool under these conditions of a greater demand. The thymidylate synthetase pathway is active in a neuroblastoma cell line and thymidylate formation is not entirely dependent on preformed thymine and thymidine kinase. It is unknown if cAMP-stimulating agents or other agents that do not change the cAMP level, all of which are known to inhibit cell division, would produce a similar change in enzyme activity.

*Ornithine Decarboxylase*

Ornithine decarboxylase (ODC) catalyzes the conversion of ornithine to the diamine, putrescine. This reaction is the rate-limiting step in the synthesis of putrescine and the naturally occurring polyamines, spermidine and spermine, which have been linked to many processes crucial for cell growth, division, and differentiation.[127,128] cAMP-stimulating agents increase ODC activity in NEI-115 neuroblastoma cells at the confluent stage, suggesting that cAMP may regulate ODC activity in neuroblastoma cells.[127] Fresh growth medium, when added to the confluent culture of neuroblastoma cells, markedly increases ODC activity. Within 15 min the degree of increase in enzyme activity is up 1000-fold. Even a relatively short pulse with fresh medium increased ODC activity. The addition of fresh medium does not change the intracellular level of cAMP. The induction of ODC by fresh serum medium is completely blocked by cycloheximide (inhibitor of protein synthesis) and actinomycin D (inhibitor of RNA synthesis), indicating the transcription of ODC has increased. Whether cAMP-induced increase in ODC activity is achieved by a similar mechanism has not been investigated. Nevertheless, it has been suggested[127] that cAMP regulates the expression of ODC at the transcription level, in analogy to its effect on the *lac* operon in *Escherichia coli*. ODC activity varies considerably during development of rat[129] and human[130] brain. Activities are high during prenatal development and low after birth. In the rat brain, greatest enzyme activities are found during cell proliferation in various regions.[129] A similar correlation between growth rate and ODC activity has been found in mouse neuroblastoma cell in culture.[131] Putrescine accumulates in the cells after the induction of ODC. The decline in ODC activity with putrescine accumulation could be interpreted in terms of feedback inhibition of ODC induction by the end product, putrescine. Indeed, the induction of ODC by fresh medium was significantly regressed when $10^{-5}$ M putrescine was added to N115 neuroblastoma cells in fresh medium.[131] The high activity of ODC in neural tumor cells may be related to their ability to proliferate. The regulation of ODC also appears to involve more than one mode, one of which is changed in the level of cAMP.

## Glucose-Metabolizing Enzymes

In neuroblastoma cells grown in serum-free medium, the activity of lactic dehydrogenase (LDH) decreases by 76%, whereas the activity of malate dehydrogenase (MDH) and glutamate dehydrogenase (GDH) increase by about 75 and 37.8%, respectively.[132] The activities of fructose diphosphate aldolase (ALD) or glucose-6-phosphate dehydrogenase (G-6PD) did not significantly change. The similarity of ALD and G-6PD activities suggests that glucose metabolism occurs at similar rates whether the cells are multiplying or differentiating. The decrease of LDH activity and the increase of MDH and GDH activities in the cells in the serum-free medium suggest that the Krebs cycle is more active in the differentiated cell. These results agree with higher oxygen uptake found in differentiating neuroblastoma cells.[133] The large increase of GDH activity may be related to a larger utilization or to a larger synthesis of one of the most important endogenous neuronal substrates, i.e., glutamic acid. A study[134] has shown that the transport of glucose in cAMP-induced (Table 3-7) differentiated cells is not significantly different from that in the malignant cells.

The electrophoretic pattern of LDH, 6-phosphogluconate dehydrogenase in Ro 20-1724- or PGE$_1$-induced differentiated cells did not change, whereas the pattern of G-6PD was markedly changed under a similar experimental condition.[8] The enzyme is expressed as a double band in Ro 20-1724-treated cells and a triple band in PGE$_1$-treated cells (Figure 3-2). However, in the absence of quantitative data the role of cAMP in the regulation of G-6PD activity in neuroblastoma cells cannot be evaluated.

MDH, GDH, and cytochrome c reductase activities are all in-

**Table 3-7.** Transport of [³H]-2-Deoxyglucose
cAMP-Induced Differentiated Mouse NB Cells
in Culture$^a$

| Treatment | % of control |
| --- | --- |
| Control (0.05 ml alcohol) | $101 \pm 11^b$ |
| PGE$_1$ (10 μg/ml, 3 days) | $73 \pm 9.4$ |
| Ro 20-1724 (200 μg/ml, 3 days) | $85 \pm 9.0$ |

$^a$ NB cells ($10^5$) were plated in 60-mm Falcon culture dish
with Ro 20-1724 (200 μg/ml), an inhibitor of cAMP phosphodiesterase activity. PGE$_1$ (10 μg/ml) was added 24 hr later.
The drug and medium were changed 2 days after treatment
and the glucose transport was performed 3 days after treatment using [³H]-2-deoxyglucose. The cells were washed
twice with warm Earl's salt and then cells were incubated
in Earl's salt containing [³H]-2-deoxyglucose (0.5 μCi/ml).
After 10 min of incubation, cells were washed twice with
cold Earl's salt, and 2 ml reagent C (used in protein determination by Lowry method) was added to each dish. An aliquot was counted for radioactivity in Toluene-Triton mixture and an aliquot was taken for protein determination.
The data were calculated as cpm/mg protein and then the
values for treated cells were expressed as % of control. Each
value represents an average of six samples. From Prasad *et
al.*[134]

volved in aerobic carbohydrate metabolism. All these enzymic activities are increased between 1 and 4 days of neuroblastoma (cell line
M$_1$) cell growth.[135] The shift of the cellular metabolism from glycolysis
to aerobiosis may represent a metabolic expression of cell differentiation, since an increase of oxygen consumption was observed in parallel.[135] Similar enzymatic modifications were observed during differentiation of neuroblastoma cells in a serum-free medium.[135]

The presence of 5-BrdU in the growth medium causes
morphological differentiation,[26] as well as increase in several enzyme
activities.[27,29] The treatment of cells at the stationary phase of 5-BrdU
further increases the activity of some enzymes.[135] 5-BrdU increases
MDH activity by 80% (39% increase without 5-BrdU), cytochrome c
reductase activity by 45% (34% increase without 5-BrdU), and GDH
activity by 42% (86% increase without 5-BrdU). At the same time, a
drastic decrease of the total LDH activity occurs. Such enzymic
modification diminishes the conversion of pyruvate to lactate and
favors coupled oxidation of glucose via the Krebs cycle, thereby increasing the oxygen uptake. In the cells approaching the stationary
phase, a decrease of LDH isoenzyme-5 and an increase of LDH
isoenzyme-3 was seen.[135] LDH-5, which constitutes only muscle-
type polypeptide subunits, is the major form in tissue exhibiting a

high rate of glycolysis and is not inhibited by pyruvate.[136] On the contrary, LDH isoenzymes that contain heart-type subunits (band 4: $M_3H$ and band 3: $M_2H_2$) are geared for activity in an aerobic metabolism. This modification in the LDH isoenzyme pattern reflects also the shift of cell metabolism from glycolysis to an aerobic metabolism, and essential phenomenon occurring during differentiation of neuroblastoma cells. In addition to causing an increase in enzyme activities involved in aerobic glucose metabolism, 5-BrdU specifically inhibits LDH synthesis.[135] The synthesis of the two types of isoenzyme subunits of LDH, muscle (M) and heart (H), is controlled by two separate loci located on different chromosomes.[136] 5-BrdU inhibits the synthesis of the M as well as of the H subunits. BrdU can replace thymidine in DNA during nucleic acid synthesis. The presence of BrdU in the DNA molecule increases the rate of pairing mistakes, since the base analogue will occasionally pair with guanine. The result is a transitional mutation that may produce an inhibition of switching locus necessary to initiate a new synthetic program. Since such changes do not occur for other enzymes, it can be suggested[135] that the incorporation of BrdU into the LDH isoenzyme gene is favored by its structure. The role of cAMP in the regulation of enzymes involved in aerobic and anaerobic glucose metabolism remains to be evaluated.

## Ganglioside-Metabolizing Enzymes

Mouse neuroblastoma clones (NB41A) contain $GM_2$, $GD_2$, $GM_1$, and $GD_{1a}$, and very little $GM_3$ ganglioside.[137] Marked variations in ganglioside patterns of various neuroblastoma clones having different electrical properties have also been observed.[138] The incubation of neuroblastoma clone (NB41) with norepinephrine (10 $\mu$M) caused a sixfold increase of the $(GM_1 + GM_2)/GM_3$ ratio in these cells. A high level of UDP-glucose:*Lac*-Tri-Cer $\beta$-galactosyltransferase was found in neuroblastoma clone (NIE-115) grown in flask or in suspension culture.[139] The activity of this enzyme was not stimulated by dibutyryl cAMP; however, the activities of two other galactosyltransferases (those catalyzing the transfer of galactose to 2-hydroxyceramide and lactosylceramide) markedly increased under a similar experimental condition.[112] A number of glycolipid:glycosyltransferases were tested in adrenergic clone (NIE-115) and cholinergic clone (NS-20) after the treatment of cells with 1 mM dibutyryl cAMP or serum-free medium.[111] There was a fivefold increase in the biosynthesis *in vitro* of the blood group-related *N*-acetylglucosamine containing tetraglycosylceramide (*Lac*-nTet-Cer) from *Lac*-Tri-cer (Glc NAc$\beta$1-3-Gal$\beta$1-4Glc-Cer) and UDP-galactose (UDP-Gal:*Lac*-Tri-Cer $\beta$-galactosyltransferase; GalT-3); however, very little activity for the biosynthesis of cerebroside (UDP-Gal:2-OH ceramide $\beta$-galactosyltransferase; GalT-1), lactosylceramide (UDP-Gal:Glc-Cer $\beta$-galactosyltransferase; GalT-2) or $GM_1$ ganglioside (UDP-Gal:$GM_2$ $\beta$-galactosyltransferase; GalT-4) was observed in NIE-115 cells treated with 1 mM dibutyryl cAMP. The stimulation of GalT-3 activity was only twofold in NS-20 clone after treatment of cells with dibutyryl cAMP. GalT-3 activity remained unchanged in cells (NIE-115 and NS-20) grown in serum-free medium. Unlike NIE-115, in NS-20 clone, GalT-2 activity was increased threefold after treatment of cells with dibutyryl cAMP, indicating that the cell response to cAMP on this criterion varies from one clone to another. The role of cAMP in regulating the activity of these enzymes cannot be fully evaluated because the criteria mentioned in the Introduction section are not met.

The lack of biosynthesis *in vitro* of higher gangliosides in NS-20 cells even after extensive neurite formation[111] is consistent with the observation that the neurite formation and the expression of many biochemically differentiated functions are not necessarily linked.[1]

Activity of ceramide galactosyltransferase in some clones of neuroblastoma (NIE-115), normal astroblast (clone NN), and glioma (C6) was nil[124,140]; however, a low activity was observed in NS-20 clone.[124] Although cerebroside sulphotransferase could not be detected in the normal astroglia cells, a notable activity was found in the glioma cells.[140] Although an increase in the sulphotransferase activity was observed in serum-free-medium-induced differentiated neuroblastoma cells, the activity of galactosyltransferase activity remains undetectable[140]; the latter enzyme, however, was inducible in the same neuroblastoma clone after treatment of cells with dibutyryl cAMP.[124] The neuroblastoma cells have lower activity of sulphotransferase than that observed in neuronal fraction of brain.[140]

## Protease Activity

The possible involvement of protease in the uncontrolled growth of transformed cell line has been suggested.[141] Human neuroblastoma cells possess a fibrinolytic activity not associated with normal human skin fibroblasts.[142] Upon incubation in a serum-free medium, the neuroblastoma cells liberate a fibrinolytic factor that requires serum for fibrin degradation. There is a strong preference for human serum. Since the serum requirement can be replaced by human plasminogen, the cell factor is probably a plasminogen activator. On the basis of sensitivity of diisopropylfluorophosphate, it was concluded that the activator is a serine protease. The unclone culture of mouse neuroblastoma produces plasminogen activator which is secreted into the growth medium. The intra- and extracellular activities of this enzyme were elevated up to 14-fold by treatment of cells with cAMP-stimulating agents.[143] PGE$_1$ and sodium butyrate were most effective followed by propionic acid and dibutyryl cAMP. Theophylline was ineffective, which is consistent with the fact that theophylline does not increase cellular cAMP in neuroblastoma cells.[144] The combination of PGE$_1$ and dibutryryl cAMP increased the enzyme activity maximally. These data suggest that cAMP may regulate the activity of plasminogen activator in neuroblastoma cells.

### Changes in Protein Synthesis in Neuroblastoma Cells during Transition from Suspension to Monolayer Culture

The transition of neuroblastoma cells (clone 41A$_3$) from suspension to monolayer growth condition involves a marked increase in the protein synthesis in a cell-free system.[145] This correlates with the increased neurite formation. If the neurite growth in monolayer is poor, the protein synthesis is not activated. It is likely that cell attachment brings about a rearrangement of the cell surface, which in turn might be the triggering signal for protein-synthesis activation. The fact that the flow of mRNA from nucleus to the cytoplasm is the same in the two growth conditions[146] suggests that the activation of protein synthesis during transition from suspension to monolayer growth condition occurs at the translational level. The longer elongation rate in lysates from monolayer cells, in spite of their higher activity in protein synthesis, might be due to a qualitative change in the mRNA population. The analysis by polyaceylanide gel electrophoresis of polypeptide chains synthesized and released in the cell-free system indicates that, at least, the relative amount of proteins synthesized by lysates from the two cell types varies; this would be in favor of some qualitative change in mRNA population during transition from suspension culture condition to monolayer.

### Changes in Ribosomal RNA in Neuroblastoma Cells during Transition from Suspension to Monolayer Culture

Ribosomal RNA synthesis and processing were investigated in suspension and monolayer cultures of neuroblastoma cells.[147] After uridine labeling, the specific activity of uridine 5'-triphosphate (UTP) and the initial rate of total RNA synthesis were the same in suspension and monolayer cells. The rate of synthesis of 45S preribosomal RNA, measured by methionine labeling, and the rate of decay of ribosomal RNA were also the same under two growth conditions. On the other hand, kinetic studies of cytoplasmic labeling showed a much lower rate of ribosomal RNA appearance in the cytoplasm of suspension cells. It has been reported[148] that during the second week of development of chick cerebral hemispheres, when morphologically undifferentiated cells are replaced by young neurons, there is an accumulation of RNA. The amount of newly synthesized ribosomal RNA per cell doubles between the 6th and 14th day of development.[149] It has been concluded that the increased amount of 28S and 18S ribosomal RNA in chick embryo brain is due to an increase in the rate of transcription. In neuroblastoma cells the higher ribosomal content in monolayer culture is due neither to an increased rate of synthesis of rRNA precursor molecules nor to a greater stability of cytoplasmic rRNA, but rather to higher rate of appearance. Whatever the mechanism of higher accumulation of rRNA content in chick embryo and neuroblastoma cells, such molecular changes apparently may be associated with differentiation of nerve cells.

Synthesis of nuclear heterogenous (Hn) RNA in suspension and monolayer culture of neuroblastoma cells has been studied by uridine labeling after inhibition of ribosomal RNA by low doses of actinomycin D.[146] Although the initial rate of HnRNA synthesis did not change in two culture conditions, accumulation of HnRNA was observed in monolayer cells after long labeling time. This may be due to a greater stability of at least some species of HnRNA species. It was concluded that higher RNA content of monolayer as compared to suspension cells is due to a greater amount of rRNA in the cytoplasm and of HnRNA molecules in the nucleus. For both RNA species, the cellular content is not regulated by their rate of synthesis but rather by the processing and/or the stability of the newly synthesized molecules.

## Changes in Total Nucleic Acid and Protein Synthesis during Differentiation

In dibutyryl cAMP-induced differentiated neuroblastoma cells [A2(1) clone], the DNA synthesis decreased by $80 \pm 3.5\%$ of control, whereas RNA and protein synthesis increased by $37 \pm 8.0$ and $19 \pm 4.1\%$, respectively.[82] Another study[150] has also found that DNA synthesis decreased, whereas protein synthesis increased in neuroblastoma cells (NB 41A) after dibutyryl cAMP treatment; however, a marked decrease in RNA synthesis was found. This discrepancy cannot be explained at this time, but the difference in the pool size of RNA precursor in two different clones of neuroblastoma cells may in part account for the difference in results from the two laboratories.

## Changes in RNA Species during cAMP-Induced Differentiation

The stimulation of morphological differentiation in uncloned cells after exposure for 48 hr to concentrations of $3 \times 10^{-7}$ to $3 \times 10^{-4}$ M papaverine or $10^{-9}$ to $10^{-3}$ M dibutyryl cAMP was associated in part with a concentration-dependent decrease in incorporation of [5-$^3$H]uridine into rRNA and HnRNA.[151] The latter effect on cellular RNA produced by papaverine occurred within 1 hr after addition to the medium, and was associated with impaired uptake of radioactive precursor into uridine nucleotides and reduction in the intracellular concentration of UTP. Dibutyryl cAMP produced a decrease in the specific radioactivity of UTP without affecting the concentration of UTP in the tumor cells. Sodium butyrate caused no significant changes in the incorporation of [5-$^3$H]uridine into rRNA and HnRNA. In contrast to the other differentiating agents examined, addition of $10^{-9}$ to $3 \times 10^{-4}$ M concentration of cAMP to the tissue culture medium enhanced the incorporation of [5-$^3$H]uridine into rRNA and HnRAN at concentrations of $10^{-4}$ M and higher. This effect was observed only at high concentrations of cyclic nucleotides and was associated with an increase in the specific activity of UTP. These studies suggest that the morphological response of neuroblastoma cells is not necessarily associated with concomitant alterations in the synthesis of RNA with agents other than cAMP. Observed changes in incorporation of [5-$^3$H]uridine into RNA appear in most instances to be due to alterations in the uptake of uridine and in pool size and specific activity of UTP.

## Changes in Poly(A)-Containing mRNA during Differentiation

After exposure of the cultures to [$^3$H]adenosine for 2 hr, the labeling of cytoplasmic RNA was much more extensive in the proliferating (control) cells than in the Ro 20-1724-treated, nondividing neuroblastoma cells. However, a greater proportion of the radioactive RNA from the Ro 20-1724-treated cells was retained by the cellulose column.[152] This suggested that in the differentiated cells there was a greater production of cytoplasmic poly(A)-containing RNA relative to total cytoplasmic RNA. When the cultures were incubated with

[$^3$H]adenosine for 18 hr, the total radioactivity within cytoplasmic RNA increased to approximately the same value in both differentiated and control cultures. In this case, the proportion of radioactive poly(A)-containing RNA was about eight times greater in the drug-treated cells than in control cells (Table 3-8). In addition to cAMP-stimulating agents, other agents are known to increase some differentiated functions. These include X irradiation, serum-free medium, and sodium butyrate.[1] Therefore, the effect of these agents on poly(A)-containing cytoplasmic RNA was investigated. X-irradiation-induced differentiated cells also show an increase in poly(A)-containing cytoplasmic RNA, but to a lesser degree (Figure 3-8). The increase in poly(A)-containing cytoplasmic RNA[153] in cAMP-induced differentiated neuroblastoma cells and X-irradiated cells is consistent with the fact that increase in the activities of many enzymes occurs under these experimental conditions. There is evidence that the enzyme induction in eukaryotic cells is preceded by poly(A) synthesis.[154] Cells treated with sodium butyrate and serum-free medium do not show any change in the amount of poly(A)-containing cytoplasmic RNA.[153] Although sodium butyrate increases the activities of some enzymes in neuroblastoma,[155] it inhibits the activites of glucokinase and hexokinase in normal liver and hepatoma cells.[156] It is possible that sodium butyrate may inhibit the activities of some enzymes in neuroblastoma cells. Thus, there may be no net increase in the cellular enzyme activities in sodium butyrate-treated neuroblastoma cells in comparison to control, and therefore there is no increase in the amount of poly(A)-containing cytoplasmic RNA.

**Table 3-8.** Incorporation of [$^3$H]Adenosine into Cytoplasmic RNA of Control Cells, and Cells Treated with Ro 20-1724[a]

| Cell treatment | Time (hr) | [$^3$H]-RNA per cell ($10^3$ count/min) | RNA that contains poly(A) (%) |
|---|---|---|---|
| Control | 2 | 10.1 ± 0.9 | 2.0 ± 0.3 |
| Ro 20-1724 | 2 | 1.65 ± 0.15[b] | 5.0 ± 0.2[b] |
| Control | 18 | 422 ± 31 | 0.3 ± 0.1 |
| Ro 20-1724 | 18 | 403 ± 48 | 2.4 ± 0.3[b] |

[a] Cells were plated in 75-cm² Falcon plastic flasks and Ro 20-1724 was added to induce differentiation. After 3 days of incubation, in the presence or absence of the drug (200 μg/ml), cells were incubated again in medium containing [$^3$H]adenosine (2 μCi/ml). Cytoplasmic RNA was then prepared from cells, and the proportion of radioactive RNA binding to cellulose columns at high ionic strength was determined. Means ± S.E.M. represent four or five independently cultured series. From Bondy et al.[152]
[b] Values significantly different in drug-treated cultures from control values ($P < 0.001$, two-tailed t-test).

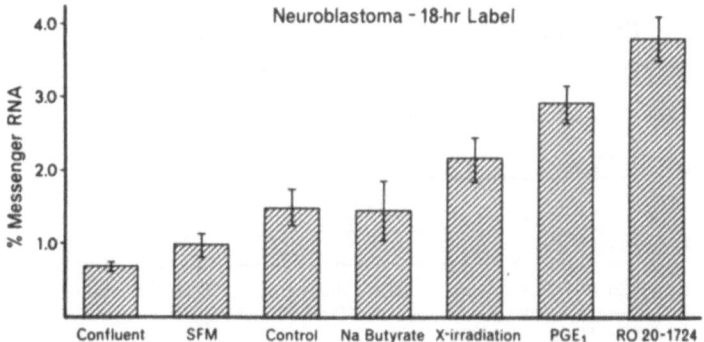

**Figure 3-8.** Proportion of labeled cytoplasmic RNA that is messenger in mouse neuroblastoma cells after various treatments. Neuroblastoma cells were plated in 75-cm² Falcon plastic flasks, and serum-free medium (SFM), sodium butyrate, (Na butyrate), PGE₁, Ro 20-1724, or X irradiation was given 24 hr later. The fresh growth medium and drug were changed every day. [³H]Adenosine (2 μCi/ml) and fresh drug were added 54 hr after treatment, and cells were further incubated for 18 hr. The polyadenylic acid containing cytoplasmic RNA in control and treated cells was determined. Each value represents an average of at least ss samples ± standard error of mean. (From Prasad *et al.*[175])

Although serum-free medium increases the activity of a few enzymes, most of the enzymes increased by cAMP-stimulating agents do not change in serum-free medium.[1] Therefore, there may not be a major increase in the cellular enzyme activities in serum-free-medium-treated neuroblastoma cells, which may be why we found no increase in the amount of poly(A)-containing cytoplasmic RNA. The lack of increase in poly(A)-containing cytoplasmic RNA in cells of confluent phase of growth is consistent with the observation that these cells do not express most of the differentiated functions induced by cAMP.[1] On the contrary, the activities of some enzymes may decrease because cells undergo degenerative changes during the confluent phase of growth.

The cytoplasm of cAMP-induced differentiated cells after 1 hr of labeling with [³H]adenosine contains threefold more radioactivity within RNA than the cytoplasm of control cells (Table 3-9). This indicates that the transport of mRNA from nucleus to cytoplasm is relatively fast in differentiated cells. However, after incubating the labeled cells for 30 hr in the absence of exogenous [³H]adenosine, the total radioactivity in RNA per cell decreased in control cells but did not change significantly in differentiated cells (Table 3-10). This indicates that the stability of total RNA in differentiated cells is greater than that in control cells. Thus, the concentration of poly(A)-

**Table 3-9.** Percentage of Total Radioactive RNA
Present within the Cytoplasms of Neuroblastoma
Cells after Various Times of Exposure to
[$^3$H]Adenosine[a]

| Labeling time | Malignant cells | Differentiated cells |
|---|---|---|
| 1 hr | 20 ± 1[b] | 57 ± 3 |
| 18 hr | 77 ± 3 | 90 ± 2 |

[a] Ro 20-1724 was used to induce differentiation. Each value represents an average of six samples. From Prasad et al. [153]
[b] S.E.M.

containing RNA and stability of total RNA in differentiated cells
increase. Therefore, cAMP may affect the differentiation of neuroblastoma cell culture by modifying the gene expression both at the translational and transcription levels. The binding of actinomycin D with
DNA suggests that the proportion of guanine residues that are acces-
sible to actinomycin D are similar to the nuclei of control and cAMP-
induced differentiated cells (Table 3-11). Thus, while the major
morphological and biochemical changes are induced by Ro 20-1724,
the quantitative amount of genetic materials available for transcription
may remain relatively constant, perhaps because some genetics are
activated, while others are suppressed during differentiation. A previous study[157] has shown that the binding of actinomycin D to
chromatin from erythroid cells does not change during maturation.

Mouse embryo and rat brain contain 6–10% of transfer RNA
phenylalanine (tRNA[phe]) species lacking peroxy-y-base; by contrast,
85% of the population from mouse neuroblastoma N18 cells are defi-
cient in the peroxy-6-nucleoside.[158] The same high percentage of

**Table 3-10.** Incorporation of [$^3$H]Adenosine into Whole
Cell RNA[a]

| Time of labeling | Malignant cells (cpm/1000 cells) | Differentiated cells (cpm/1000 cells) |
|---|---|---|
| 18 hr | 320 ± 20[b] | 870 ± 10 |
| 18 hr plus 30 hr in medium without exogenous radioisotope | 220 ± 20 | 960 ± 10 |

[a] Ro 20-1724 was used to induce differentiation. Each value represents an average of six samples. From Prasad et al. [153]
[b] S.E.M.

**Table 3-11.** Binding of [³H]Actinomycin D with Nuclei of Malignant and Differentiated Neuroblastoma Cells[a]

| Status of cells | Bound cpm/mg DNA |
| --- | --- |
| Malignant | 7,200 ± 460[b] |
| Differentiated | 7,930 ± 580 |
| Calf thymus DNA | 84,000 |

[a] Ro 20-1724 was used to induce differentiation. From Prasad *et al.* [153]
[b] S.E.M.

peroxy-y-deficient tRNA[phe] was observed in dimethylsulfoxide-induced differentiated N18 cells. Thus mouse neuroblastoma cells differ markedly from normal brain cells in this respect.

## Changes in Synthesis and Phosphorylation of Histone and Nonhistone Proteins during Differentiation

The synthesis of histone and phosphorylation of $H_1$-histone are markedly reduced (Figures 3-9 and 3-10) in cAMP-induced differentiated neuroblastoma (NBA$_{2(1)}$ clone) cells[78] which did not divide. Since the synthesis of $H_1$-histone is linked with cell proliferation,[159-162] it has been suggested that a reduction in $H_1$-histone syn-

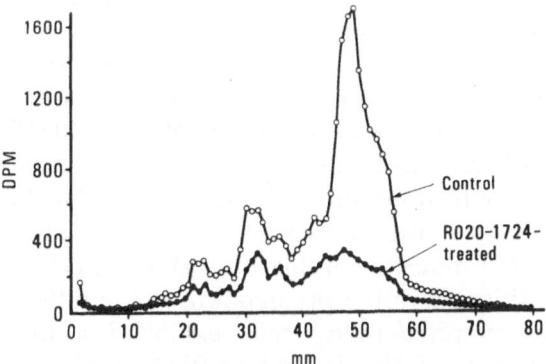

**Figure 3-9.** Distribution of radiolabeled amino acids in SDS polyacrylamide gels of various histone fractions (0.25 N HC1 fraction) from control and cAMP-induced differentiated mouse neuroblastoma cells. Cells were treated with PGE₁ or Ro 20-1724 for 3 days. Fresh growth medium was added 1 hr before the addition of radioactive amino acids. Control and treated cells were labeled with [¹⁴C] or [³H]amino acids for 2 hr. Total DPM ¹⁴C/25,133; ³H/8975. Gels were run for 17.5 hr at 6 mA/gel. (From Lazo *et al.* [78])

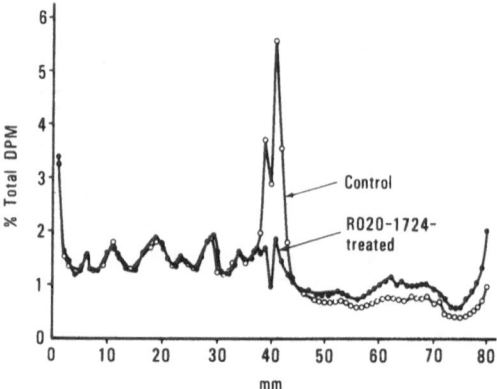

**Figure 3-10.** Distribution of radiolabeled phosphate in SDS polyacrylamide gels of various histone fractions (0.25 N HCl fraction) from control and cAMP-induced differentiated mouse neuroblastoma cells. Cells were treated with $PGE_1$ or Ro 20-1724 for 3 days. Fresh growth medium was added 1 hr before the addition of radioactive phosphate. Control and treated cells were labeled with $^{32}P$ or $^{33}P$ for 2 hr. Total DPM $^{32}P/298,079$; $^{33}P/108,670$. Gels were run for 21 hr at 6 mA/gel. (From Lazo *et al.*[78])

thesis in dividing neuroblasts may be an important biological signal to turn off cell division. However, the synthesis and phosphorylation of nonchromosomal proteins changed only slightly.[78] Other investigators[163-165] have been unable to demonstrate any significant changes in nonhistone chromosomal proteins during development of rat brains. However, we have observed a small decrease in the synthesis, and a small increase in phosphorylation, of 40,000-dalton peptides in cAMP-induced differentiated cells. The significance of this change in nonhistone protein is unknown. It has been suggested[166] that a nonhistone protein of 40,000- to 50,000-dalton range may be involved in DNA replication in mammalian cells. If this is true, the changes in the synthesis and phosphorylation of 40,000-dalton peptides in the differentiated cells may be a reflection of inhibition of DNA synthesis which occurs in these cells.[82] It has been shown[167] that modifications in the composition and transcriptional properties of the genome in mouse neuroblastoma (NB-2a) cells are associated with the transition from the proliferating to the differentiated state. Chromatin from proliferating cells exhibited a threefold-greater template activity for DNA-dependent RNA synthesis in a cell-free system than chromatin from serum-free-medium-induced differentiated cells. The electrophoretic profiles show that the five principal histone fractions represented in the nuclei of proliferating and serum-free-medium-induced differentiated cells were similar; the relative amount of protein in each fraction is also similar. However, the synthesis of nonhistone chromosomal proteins that migrate in the 120,000- to 130,000-molecular-weight regions of the gel decreased by a factor of three in both serum-free-medium- and dibutyryl-cAMP-induced differentiated neuroblastoma cells.[167] It has been noted[78] that only a slight decrease in this fraction occurs. Therefore, the significance of nonhistone chromosomal proteins during differentiation of neuroblastoma cells remains to be evaluated.

## Changes in Tubulin Synthesis during Differentiation

Microtubules are found in all eukaryotic cells, but higher concentrations of microtubule proteins are found in the brain and particularly in the axons and dendrites of the central nervous system. Microtubules are composed of an acidic protein called tubulin, with a molecular weight of about 110,000. Tubulin in itself is a dimer of two different polypeptides having molecular weights of approximately 54,000 and 56,000. Microfilaments are another important component of the axon, and at least some microfilaments are thought to contain actin. In addition actin may be associated with myosin at the synaptic junction.

Mouse neuroblastoma cells contain microtubular proteins.[168] High concentrations of glycerol (5.0 M) and sucrose (1.2 M) increase the stability of tubulin.[169] A heterologous cell-free system, directed with polyribosomes from mouse neuroblastoma cells (NS-20) and containing cell extract from rat brain, is capable of synthesizing proteins *in vitro*.[170] The extent of *in vitro* amino acid incorporation was found to depend greatly on the quality of polyribosome preparation. Highest incorporation was observed with preparations exhibiting a high ratio of polyribosomes to 80 S ribosomes. The maximum incorporation requires the presence of the microsomal washed fraction that is supposed to contain initiating factor. Thus, both chain initiation and completion of existing polypeptide chains occurred in the complete *in vitro* system. Tubulin accounted for 2% of total protein synthetized. Another study[171] also reported *in vitro* translation of tubulin and actin with mRNA from rat brain in a system free of wheat germ cells.

Tubulin from both mouse and human neuroblastoma cells has a molecular weight similar to that obtained from mouse or rat brain, but the relative amount of tubulin in the adult brain is twice as high as in neuroblastoma extract.[141] The low level of tubulin in neuroblastoma cells is associated with the low level of corresponding mRNA.

The relative level of tubulin in exponential cells was similar to that observed in stationary neuroblastoma cells (NIE-115 clone). The level remains unchanged in serum-free-medium- and aminopterin-induced morphologically differentiated cells.[171] This indicates that the preexisting pool of tubulin is utilized for neurite formation. However, a small but consistent increase in tubulin level was found in NIE-115 cells, which form neurites after DMSO treatment; but no such effect was observed in human neuroblastoma cells (N-I clone), which do not form neurite. Since the level of tubulin in both mouse and human neuroblastoma was similar, the lack of response of human cell line after DMSO treatment cannot be due to low level of tubulin.

Myosin also has been isolated from clonal line of mouse neuroblastoma ($N_2A$) and rat glioma (C6). Myosin represented 0.1–1.5% of total cell proteins and is similar in quantities and biochemical properties in both monolayer and suspension neuroblastoma cultures.[172] Actin has also been found in mouse neuroblastoma cells.[173] The relative level of actin in all neuroblastoma clones extract was similar to that in adult rat or mouse brain, but significantly lower than that in newborn brain cells.[171] The percent of actin did not change in DMSO-induced differentiated cells. On the other hand, the actin-containing fibers are markedly increased in Ro 20-1724-induced differentiated neuroblastoma cells ($P_2$ clone) compared to their controls.[174] This system may be useful in examining the regulation of actin filament assembly and its relationship to cellular morphogenesis.

### Changes in Nervous-System-Specific Protein (14-3-2) during Differentiation

The nervous-system-specific protein (14-3-2) is present in human neuroblastoma line IMR-32[47] and mouse neuroblastoma clone (NB 41A3). The relative amount of this protein is constant in dividing and stationary phase in both human and mouse neuroblastoma cells and in cells grown in monolayer or suspension culture.[175]

### Changes in the Cyclic Nucleotide System during Differentiation

A marked change in the cyclic nucleotide system occurs during differentiation of neuroblastoma cells. A diagrammatic representation of the metabolism of cyclic nucleotide is shown in Figure 3-11. We will discuss changes in each parameter of the cyclic nucleotide system during differentiation.

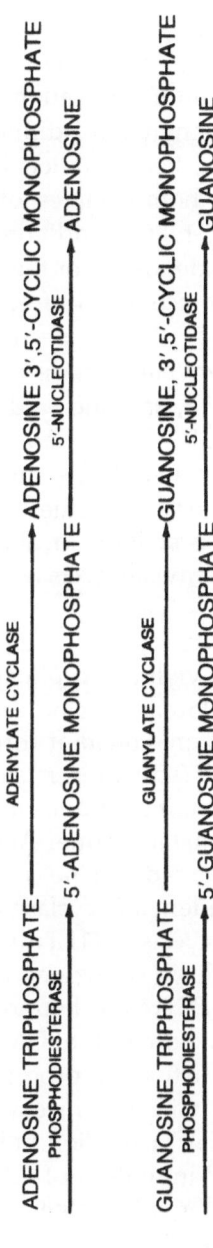

**Figure 3-11.** A diagrammatic representation of metabolism of cAMP and cGMP in mammalian cells.

*Adenylate Cyclase*

The adenylate cyclase activity in homogenates of neuroblastoma cells can be stimulated by PGE$_1$[48,176,177]; DA and NE[48,176]; apomorphine, epinephrine, sodium fluoride[48]; adenosine[176,178]; ACh only in certain clones[48,179]; and guanylylimid diphosphate.[180] The regulation of adenylate cyclase has been studied in detail.[177,178,180] The adenylate cyclase activity in homogenates of neuroblastoma cells (NS-20) was increased 2.5- to 4-fold by addition of 200 $\mu$M adenosine, 2-chloroadenosine, 2-hydroxyadenosine, or 8-methylaminoadenosine. The $K_m$ of homogenate cyclase activation of adenosine and 2-chloroadenosine was 67.6 and 6.7 $\mu$M, respectively. Two classes of inhibitors of homogenate adenylate cyclase activity were observed. One class, which included AMP, adenine, and theophylline, blocked 2-chloroadenosine but not PGE$_1$ stimulation of cyclase. The second class of inhibitors, which included 2'- and 5'-deoxyadenosine, inhibited unstimulated 2-chloroadenosine- and PGE$_1$-stimulated homogenate adenylate cyclase activity to about the same degree. Activation of homogenate adenylate cyclase by adenosine appears to be noncooperative.

The process of adenylate cyclase stimulation of PGE$_1$, 2-chloroadenosine, and 5'-guanylylimid diphosphate [GMP-P(NH)P] is different.[180] PGE$_1$ and 2-chloroadenosine activation is rapid, producing elevated activities that are constant throughout a 20-min assay. In contrast, GMP-P(NH)P activation is slow, and although the activity is elevated within 1 min, it continues to increase up to 12 min before attaining a maximal constant value. Activation is more rapid when either PGE$_1$ or 2-chloroadenosine is present with GMP-P(NH)P. The neuroblastoma adenylate cyclase has the following regulation properties common to cyclases: (1) a nucleotide regulatory site specific for guanine nucleotides; (2) enzyme activity that is greatly increased when GTP of GMP-P(NH)P is bound at this site and that is very low in their absence; (3) a more potent GMP-P(NH)P, than GTP, activation; (4) a GMP-P(NH)P activation process that is "irreversible" and, therefore, $E_{GMP-P(NH)P}$ enzyme not in equilibrium with the $E_{(0)}$ enzyme state; (5) a GMP-P(NH)P activation inhibited by low concentrations of GDP; and (6) GDP inhibition of GMP-P(NH)P activation regulatable by PGE$_1$ and chloroadenosine, two nonguanine nucleotide activators of the cyclase. Therefore, it has been proposed[180] that GTP, like GMP-P(NH)P, is "irreversible" bound to the enzyme; that GTP, like GMP-P(PH)P, is rapidly converted to GDP; and

that GDP is tightly bound to the enzyme but dissociable. It follows
that adenylate cyclases revolve through a cycle,

$$E_{(0)} \overset{(GTP)}{\rightarrow} E_{(GTP)} \rightarrow E_{(GDP)} \overset{\leftarrow}{\rightarrow} E_{(0)}$$

and that most of an "unstimulated" enzyme population is in the $E_{(GDP)}$
state. By shifting the equilibrium $E_{(GDP)} \leftrightarrows E_{(0)}$ toward $E_{(0)}$, $PGE_1$,
chloroadenosine, and other possible "secondary" activators of
adenylate cyclase would make more of the enzyme available for GTP
binding and would thereby increase the amount of the active enzyme
(i.e., $E_{(GTP)}$). It has been reported[176] that $PGE_1$ was a much more
potent stimulatory agent on adenylate cyclase activity than
adenosine. The reason why clear saturation kinetics were not obtained when concentrations of $PGE_1$ were increased to values higher
than $10^{-5}$ M is not yet known. One possible reason is that the strongly
hydrophobic $PGE_1$ could exert a detergentlike effect at high concentrations. It has been observed[181] that low concentrations of detergents
were able to increase adenylate cyclase activity. Therefore, the pronounced stimulatory effect of $PGE_1$ lower than $10^{-6}$ M reflects the
existence of specific receptors.

In contrast to the result of Blume et al.,[131,178,180] analogs of
adenosine 2-deoxyadenosine, 3-deoxyadenosine, and adenine $\beta$-
D-arabinofuranside did not activate adenylate cyclase activity, although adenosine did.[176] 5'-AMP and adenosine at concentrations
higher than $10^{-4}$ M inhibit adenylate cyclase activity, which has been
interpreted as an interaction with catalytic moiety. This may explain
the inhibitory effect of adenosine ($10^{-4}$ M) on $PGE_1$-stimulated adenylate cyclase activity. Although theophylline antagonizes the stimulatory effect of adenosine on cyclase activity by reducing both the apparent $K_m$ and $V_{max}$, it does not reverse the inhibitory effect of adenosine
on $PGE_1$-stimulated adenylate cyclase activity. Thus, depending
upon the concentration, adenosine may stimulate or inhibit the enzyme activity. The nature of adenosine receptor is unknown. Mouse
neuroblastoma adenylate cyclase has many properties in common
with the enzyme obtained from nonneural tissue. As found in other
adenylate cyclase, the neuroblastoma enzyme requires a divalent cation such as $Mg^{2+}$ or $Mn^{2+}$ in addition to ATP for activity[47,177] It has
been postulated that the true substrate for all adenylate cyclases is a
Mg–ATP complex, and that the stimulation of the $Mg^{2+}$ level in excess
of the ATP levels is due to the ion's continued chelation of the free

ATP, which acts as a cyclase inhibitor. High concentrations of $Mg^{2+}$ and $Mn^{2+}$ (15 mM) inhibited adenylate cyclase activity; the effect was more pronounced with $Mn^{2+}$.[49] Other investigators[177] also noted inhibition of enzyme activity by $Mn^{2+}$. A 3-mM concentration of calcium increased adenylate cyclase activity 50% of control; a higher concentration of 10 mM did not significantly affect the enzyme activity in control cells, but slightly inhibited (29%) it in cAMP-induced differentiated neuroblastoma cells.[49] It has been shown[177] that calcium inhibits basal level and 2-chloroadenosine- and $PGE_1$-stimulated adenylate cyclase activity in NB cells in a cooperative manner.

## Pharmacological Characterization of Adenylate Cyclase

Haloperidol, which is known to block the dopamine (DA) receptor, inhibited DA-stimulated adenylate cyclase acivity; however, it required a much higher concentration for a similar inhibition of NE-stimulated enzyme activity (Table 3-12). A concentration of 11 $\mu M$ haloperidol reduced DA-stimulated enzyme activity by 50%; however, even a high concentration (100 $\mu M$) of haloperidol decreased NE-stimulated enzyme activity by about only 30%. Haloperidol produced no significant effect on nonstimulated adenylate cyclase activity.

About 0.43 $\mu M$ phentolamine (blocker of $\alpha$ receptors) reduced DA-stimulated adenylate cyclase activity by 50%, whereas a similar amount of inhibition of NE-stimulated enzyme activity was achieved with about 2.3 $\mu M$ phentolamine (Table 3-12). About 180 $\mu M$ propranolol (blocks $\beta$ receptors) produced a 50% inhibition of DA-

**Table 3-12.** Effect of Blocking Agents on Catecholamine-Stimulated Adenylate Cyclase Activity in Homogenates of Neuroblastoma Cells[a]

|  | Concentration ($\mu M$) which produces 50% inhibition | | |
|---|---|---|---|
|  | Haloperidol | Phentolamine | Propranolol |
| DA-stimulated adenylate cyclase activity | 11 | 0.43 | 180 |
| NE-stimulated adenylate cyclase activity | >100 | 2.3 | 1.7 |

[a] Effect of haloperidol, phentolamine, and propranolol on DA-and NE-stimulated adenylate cyclase activity in homogenates of differentiated mouse neuroblastoma cells. Ro 20-1724 was used to induce differentiation. The adenylate cyclase activity in homogenates was measured 3 days after treatment with Ro 20-1724. From the curves of each blocking agent the concentration that inhibited 50% of the catecholamine-stimulated adenylate cyclase activity was determined. Each value represents an average of six samples. From Prasad and Gilmer.[48]

stimulated adenylate cyclase activity, whereas NE-stimulated enzyme activity required only 1.7 $\mu$M propranolol for a similar amount of inhibition (Table 3-12). Neither $\alpha$-adrenergic nor $\beta$-adrenergic blocking agents affected the basal activity of adenylate cyclase. These data suggest that the DA-sensitive adenylate cyclase has pharmacological properties different from those of NE-sensitive adenylate cyclase. This is further supported by the following observations: (1) DA and NE produce an additive stimulatory effect on adenylate cyclase activity (Table 3-13); and (2) the combination of $PGE_1$ and NE produces an additive stimulatory effect on adenylate cyclase activity, whereas the combination of $PGE_1$ and DA does not.

The pharmacological properties of DA receptors are similar to those of $PGE_1$ receptors. For example, $PGE_1$-stimulated adenylate cyclase is blocked by a low concentration of phentolamine and haloperidol (Table 3-13). The combination of DA and $PGE_1$ does not produce an additive stimulatory effect on adenylate cyclase activity.

ACh-sensitive adenylate cyclase is not demonstrable in one clone ($NBA_{2(1)}$) of mouse neuroblastoma cells[48] but is detectable in another clone ($NBDB^-$) (Figure 3-12). ACh-stimulated adenylate cyclase activity is markedly blocked by inhibitors of muscarinic receptors (atropine) and by inhibitors of nicotinic receptors (nicotine and hexamethonium); however, ACh-stimulated adenylate cyclase activity is not significantly affected by haloperidol, propranolol, and phentolamine.[179] ACh, DA, and NE produced an additive stimulatory ef-

Table 3-13. Effects of Prostaglandin and Catecholamines on Adenylate Cyclase Activity in Homogenates of Differentiated Neuroblastoma Cells[a]

| Treatment | Adenylate cyclase activity (pmol/mg protein/min) |
|---|---|
| Basal Level | $21 \pm 1^b$ |
| DA (100 $\mu$M) | $36 \pm 4$ |
| NE (100 $\mu$M) | $42 \pm 5.2$ |
| $PGE_1$ (10 $\mu$M) | $41 \pm 1$ |
| DA + NE | $69 \pm 3$ |
| NE + $PGE_1$ | $74 \pm 2$ |
| DA + $PGE_1$ | $38 \pm 1$ |
| $PGE_1$ + propranolol (10 $\mu$M) | $40 \pm 2$ |
| $PGE_1$ + phentolamine (10 $\mu$M) | $20 \pm 1$ |

[a] Ro 20-1724 was used to induce differentiation. Each value represents an average of six samples. From Prasad and Gilmer.[48]
[b] S.D.

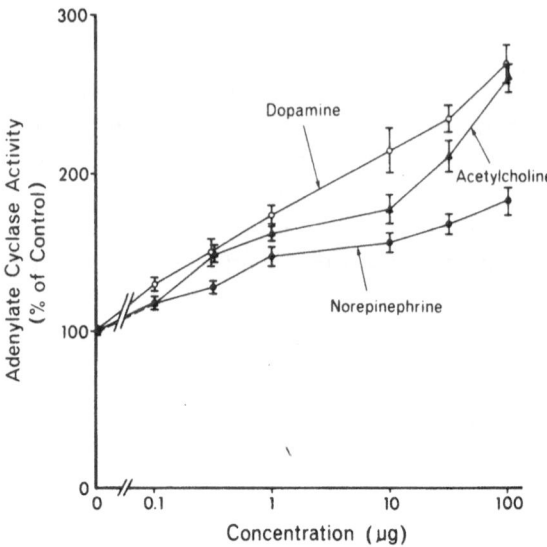

**Figure 3-12.** Changes in adenylate cyclase activity (AC) in homogenates of inactive neuroblastoma cells (NBDB⁻). The basal activity of AC (22 ± 2 pmol/mg protein/min) was considered 100% control values, and the AC values of treated cells were expressed as % of control. Each value represents an average of four to six samples. The bar at each point is standard deviation. Treatment with $DA_1$, ACh, and NE. (From Prasad et al. [179])

fect on adenylate cyclase activity (Table 3-14), suggesting that these neurotransmitters act on different receptor sites, all of which are linked with adenylate cyclase. The presence of ACh receptors on neuroblastoma cells was first demonstrated by Harris and Dennis.[183] These ACh receptors are blocked by nicotinic and muscarinic inhibitors.[105]

## Changes in Adenylate Cyclase Activity during Differentiation

In cAMP-induced differentiated cells, the DA concentration needed for a maximal increase in enzyme activity is 10 times less than that needed in control cells (Figure 3-13). In addition, NE-sensitive adenylate cyclase activity becomes detectable at low NE concentration.[48] However, in homogenates obtained from serum-free-medium- and X-irradiation-induced differentiated cells, DA and NE failed to stimulate adenylate cyclase activity.[48,49] On the other hand, the adenylate cyclase activity in homogenates obtained from sodium butyrate-treated neuroblastoma cells was higher than that obtained from control cells (Table 3-15); the enzyme activity was further stimulated by DA and NE.[48]

**Table 3-14.**   Effect of Combination of Various
Neurotransmitters on Adenylate Cyclase
Activity in Homogenates of Neuroblastoma
Cells (NBDB $^-$)$^a$

| Treatment | Adenylate cyclase activity (pmol/mg protein/min) |
|---|---|
| Control | 22 ± 2$^b$ |
| DA (100 $\mu$M) | 58 ± 3 |
| NE (100 $\mu$M) | 40 ± 1 |
| Epinephrine (100 $\mu$M) | 26 ± 2 |
| ACh (100 $\mu$M) | 57 ± 4 |
| Bethanechol (100 $\mu$M) | 36 ± 2 |
| Carbachol (100 $\mu$M) | 32 ± 2 |
| ACh + bethanechol | 88 ± 3 |
| ACh + carbachol | 77 ± 2 |
| DA + ACh | 100 ± 2 |
| NE + ACh | 92 ± 5 |
| DA + NE + ACh | 133 ± 3 |
| Serotonin (100 $\mu$M) | 24 ± 2 |
| Histamine (100 $\mu$M) | 24 ± 2 |
| PGE$_1$ (10 $\mu$M) | 52 ± 3 |

$^a$ From Prasad et al. [179]
$^b$ S.D.

The effect of GTP on adenylate cyclase activity was similar in control and cAMP-induced differentiated cells.[49] A concentration of 10 $\mu$M GTP stimulated adenylate cyclase activity by 35%. PGE$_1$-stimulated adenylate cyclase activity was similar in both control and cAMP-induced differentiated neuroblastoma cells (Figure 3-14). The combination of PGE$_1$ and GTP (10 $\mu$M) did not produce a significant change in adenylate cyclase activity of control cells when compared to the effect of PGE$_1$ alone. However, in cAMP-induced differentiated neuroblastoma cells, GTP stimulated the PGE$_1$ effect on adenylate cyclase activity at all PGE$_1$ concentrations.[49]

A concentration of 3 mM calcium stimulated adenylate cyclase activity by about 50% in homogenates of control and cAMP-induced differentiated neuroblastoma cells. At a higher concentration (10 mM), calcium had no significant effect on adenylate cyclase activity of control cells, but it inhibited the enzyme activity of differentiated cells by about 20%.[49] A concentration of 15 mM Mg$^{2+}$ inhibited adenylate cyclase activity in control and differentiated neuroblastoma cells by about 20 and 35%, respectively. The effect of manganese on adenylate cyclase activity was more pronounced than that of magnesium. A

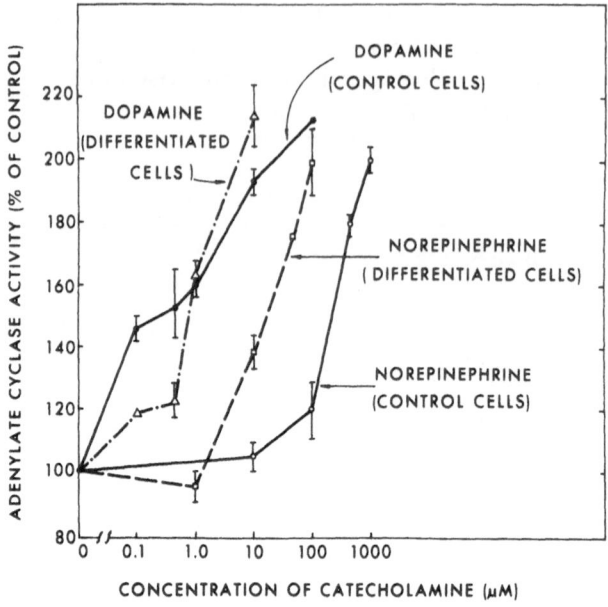

**Figure 3-13.** Changes in adenylate cyclase activity in homogenates of control and differentiated mouse neuroblastoma cells NBA$_{2(1)}$ after treatment with DA and NE. Ro 20-1724 was used to induce differentiation. The basal activities of adenylate cyclase in control (15 ± 1.4 pmol/mg protein/min) and differentiated (21 ± 1 pmol/mg protein/min) cells were considered 100% control values; the adenylate cyclase values of treated cells were expressed as percentage of control. Each value represents an average of 8–12 samples. The bar at each point is standard deviation.(From Prasad and Gilmer.[48])

**Table 3-15.** Effect of DA and NE on Adenylate Cyclase Activity in Homogenates of Serum-Free-Medium and Sodium-Butyrate-Treated Neuroblastoma Cells[a]

|  | Adenylate cyclase activity (pmol/mg protein/min) | | |
| --- | --- | --- | --- |
| Treatment | Basal | DA (100 μM) | NE (100 μM) |
| Control | 15 ± 1.4 | 32 ± 1 | 18 ± 2.2 |
| Sodium butyrate (0.5 mM) (3 days) | 35 ± 1.4 | 60 ± 1.8 | 68 ± 1.5 |
| Serum-free medium (3 days) | 16 ± 1.5 | 19 ± 1.2 | 17 ± 1 |

[a] The adenylate cyclase activity in homogenates was measured 3 days after treatment. Each value represents an average of six samples ± S.D. From Prasad and Gilmer.[48]

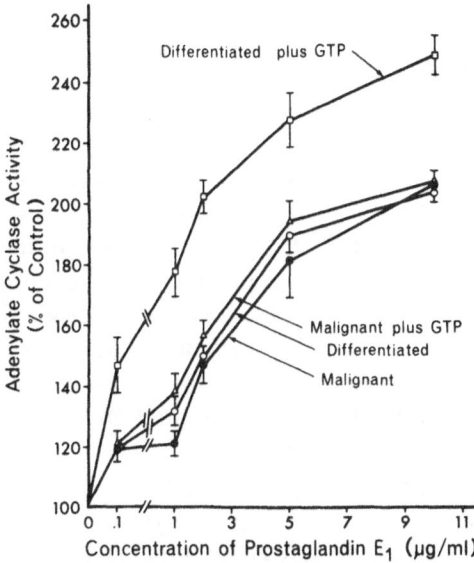

**Figure 3-14.** Changes in adenylate cyclase activity in homogenates of control (malignant) and differentiated mouse neuroblastoma cells ($NBP_2$) after treatment with $PGE_1$ and guanosine triphosphate (10 $\mu$M). Ro 20-1724 was used to induce differentiation. The basal activities of adenylate cyclase in malignant (23 ± 1 pmol/mg protein) and differentiated (25 ± 1 pmol/mg protein/min) cells were considered 100% control values; the adenylate cyclase values of treated cells were expressed as percentage of control. Each value represents an average of six samples. The bar at each point is standard deviation. (Prasad *et al.*[49])

concentration of 10 mM manganese inhibited adenylate cyclase activity in control and differentiated cells by about 29 and 46%, respectively.[49]

There was no statistically significant change in the value of $K_m$ or $V_{max}$ of adenylate cyclase in differentiated cells when compared to control cells.[49]

These data show that the sensitivity of adenylate cyclase to neurotransmitters and divalent ions, and the sensitivity of $PGE_1$-stimulated enzyme activity to GTP increase in differentiated neuroblastoma. These changes in neuroblastoma cells could not be expressed because of malignant transformation of dividing neuroblasts.

## Guanylate Cyclase Activity

The regulation of guanylate cyclase activity of neuroblastoma cells in control or differentiated cells has not been studied.

## Regulation of Intracellular Level of cAMP

The intracellular level of cAMP can be affected by relative activity of adenylate cyclase, sensitivity of adenylate cyclase to neurotransmitter, hormones and ions, activity of phosphodiesterase, and the level of cAMP binding protein. Several agents increase the intracellular level of cAMP. Among stimulators of adenylate cyclase, $PGE_1$[81,184] and adenosine[185,186] increase the intracellular level of cAMP in several neuroblastoma clones. Among inhibitors of cyclic nucleotide phosphodiesterase, Ro 20-1724 increases the intracellular level of cAMP, but theophylline does not.[81] In some clones, $PGE_1$ or Ro 20-1724 by itself fails to increase the intracellular level of cAMP (Table 3-16); but when they are combined the intracellular level of cAMP increased to a level which is produced in a sensitive clone under a similar experimental condition.[187] Although DA and NE stimulate adenylate cyclase activity *in vitro* in some clones of neuroblastoma cells,[48] they do not increase the cAMP level until the phosphodiesterase activity is reduced by Ro 20-1724.[134] Thus it appears that at least in some clones of neuroblastoma, the activity of cAMP phosphodiesterase is the rate-limiting factor in the accumulation of cAMP after treatment of neuroblastoma cells with DA, NE, $PGE_1$, and adenosine. Previous studies[184,185] have failed to observe the stimulatory effect of DA and NE on the cAMP level in the present or absence of a phosphodiesterase inhibitor. This would be because these investigators studied DA response on the cAMP level in serum-free

**Table 3-16.** Effect of Neurotransmitters and $PGE_1$ on cAMP Level in Neuroblastoma Clones[a]

| Treatment | cAMP levels in various clones (pmol/mg protein) | | | |
| --- | --- | --- | --- | --- |
| | $NBA_{2(1)}$ | $NBE^-(A)$ | $NBDB^-$ | $NBP_2$ |
| Control | 14 ± 2[b] | 19 ± 2 | 19 ± 4 | 11 ± 2 |
| Ro 20-1724 | 48 ± 7 | 107 ± 5 | 26 ± 4 | 65 ± 8 |
| $PGE_1$ | 20 ± 2 | 22 ± 22 | 56 ± 9 | 19 ± 2 |
| Ro 20-1724 + DA | 233 ± 14 | 139 ± 4 | 162 ± 49 | 66 ± 5 |
| Ro 20-1724 + NE | 191 ± 19 | 120 ± 11 | 122 ± 36 | 67 ± 4 |
| Ro 20-1724 + ACh | 16 ± 2 | 47 ± 7 | 19 ± 4 | 23 ± 4 |
| Ro 20-1724 + bethanechol | 37 ± 2 | 101 ± 16 | 26 ± 2 | 52 ± 7 |
| Ro 20-1724 + $PGE_1$ | 2803 ± 173 | 1953 ± 118 | 2426 ± 124 | 2404 ± 212 |

[a] $NBA_{2(1)}$ contains TH, but not CAT; $NBE^-(A)$ contains CAT but no TH; $NBDB^-$ contains neither TH nor CAT; $NBP_2$ contains both CAT and TH. From Sahu and Prasad.[187]
[b] S.D.

medium; it has been shown[48] that DA does not stimulate adenylate cyclase activity in homogenates when cells are treated with serum-free medium. Another reason could be the difference in the clones used by investigators. Indeed, we have observed[187] that DA and NE failed to increase cAMP level even in the presence of a phosphodies-terase inhibitor in some clones. This shows that many cells in neuroblastoma tumor lack DA and NE receptors. Isoproterenol stimulates adenylate cyclase activity in homogenates of untreated neuroblastoma cells (clone $NBA_{2(1)}$),[48] but it does not increase the cAMP level in the intact cells either in the presence or absence of a phosphodies-terase inhibitor. Serotonin and histamine neither stimulate adenylate cyclase activity in homogenates nor elevate the cAMP level in the intact cells in the presence or absence of a phosphodiesterase inhibitor.[186]

The effect of various neurotransmitters on the accumulation of cAMP level in various clones of nerve and glial cells (derived from chemically induced CNS tumor) were studied.[188] Two of the five neuronal lines responded to NE with an increase in endogenous cAMP, while 9 of the 11 glial lines responded; the response was larger in the majority of glial lines than in the nerve cells. DA response was seen in glial cells, but not in nerve cells. The failure to observe the effect of DA and NE may be due to high activity of cAMP phos-phodiesterase in these clones. Therefore, until the responses of these drugs were studied in the presence of a phosphodiesterase inhibitor, the lack of DA and NE responses in some clones of glial or in all clones of nerve cells cannot be fully evaluated. The catecholamine response including the DA response in responsive clone was mediated through $\beta$ receptors.[188]

Although ACh has no effect on adenylate cyclase activity *in vitro* in one neuroblastoma clone ($NBA_{2(1)}$),[48] it stimulates the enzyme activity in other clones.[179] However, ACh decreases the intracellular level of cAMP raised by phosphodiesterase inhibitor in all clones.[186] ACh decreases the intracellular level of cAMP augmented by $PGE_1$ and adenosine[189,190] and by DA, NE, $PGE_1$, and adenosine in the presence of phosphodiesterase inhibitor (Table 3-17). The effect of ACh is observed in a medium with *minimum* calcium or magnesium, and also in a medium *free* of calcium or magnesium, indicating that these ions are not important in producing the ACh effect.[1] The effect is also not due to leakage of formed cAMP into the medium, since the amount of cAMP in the medium of culture treated with ACh and Ro 20-1724 was similar to that observed in the medium of Ro 20-1724-treated cells.[1] The ACh effect is primarily mediated via muscarinic

Table 3-17.  Effect of ACh on DA-, NE-, Adenosine-, and PGE$_1$-Increased Level of cAMP in Neuroblastoma Cells (A$_{2(1)}$)[a]

| Treatment | cAMP level (% of increased level) |
|---|---|
| Ro 20-1724 + ACh | 49 ± 4[b] |
| Ro 20-1724 + DA + ACh | 53 ± 8 |
| Ro 20-1724 + NE + ACh | 36 ± 7 |
| Ro 20-1724 + adenosine + ACh | 47 ± 2 |
| Ro 20-1724 + PGE$_1$ + ACh | 50 ± 5 |
| Ro 20-1724 + bethanechol | 105 ± 9 |

[a] The control unstimulated values of cAMP vary from 25 to 35 pmol/mg protein. DA and NE in the presence of Ro 20-1724 increased the intracellular level of cAMP by about 20-fold. Adenosine and PGE$_1$ in the presence of Ro 20-1724 increased the cAMP level by about 180- and 150-fold, respectively. From Prasad.[1]

receptors, because it is blocked by atropine and hexamethonium but not by nicotine (Table 3-18). In addition, bethanecol, which has primarily a nicotinic effect, does not mimic the ACh effect. Hence, the ACh-induced decrease in the intracellular level of cAMP may be due to a decrease in muscarinic-linked adenylate cyclase activity. However, in vitro study[179] has shown that ACh stimulates the activities of both muscarinic- and nicotinic-linked adenylate cyclases. These data show that the regulation of muscarinic- and nicotinic-linked adenylate cyclase activity in vivo is different from that observed in vitro. The

Table 3-18.  Effect of Inhibitors of ACh Receptors on the ACh Effect on cAMP in Neuroblastoma Cells (NBP$_2$)[a]

| Treatment | cAMP level (pmol/mg protein/min) |
|---|---|
| Control | 5 ± 0.4[b] |
| Ro 20-1724 (200 μg/ml) | 29 ± 0.4 |
| ACh (100 μM) | 6 ± 0.5 |
| Ro 20-1724 + bethanechol (100 μM) | 33 ± 3.0 |
| Ro 20-1724 + carbachol (100 μM) | 6 ± 0.5 |
| Ro 20-1724 + ACh + atropine (100 μM) | 28 ± 1.4 |
| Ro 20-1724 + ACh + hexamethonium (30 μM) | 22 ± 1.2 |
| Ro 20-1724 + ACh + nicotine (10 μM) | 8.7 ± 0.7 |

[a] Cells (0.5 × 10⁶) were plated in large 75-cm² Falcon plastic flasks. At 3 days after plating, fresh medium was substituted and then Ro 20-1724, a specific inhibitor of cAMP phosphodiesterase, was added. After 1 hr of incubation, ACh alone or in combination with one of the inhibitors of ACh receptors was added. Cells were incubated in the presence of drugs for an additional 15 min, and the cAMP level was determined. Each value represents an average of twelve to fourteen samples. From Prasad.[1]
[b] S.E.M.

level of cAMP in neuroblastoma cells increases within 5 min of treatment of cells with PGE₁ and Ro 20-1724 and remains high for a period of observation (Figure 3-15). Sodium butyrate[155] and 5-BrdU[26,27,29] are known to induce several differentiated functions in mouse and human neuroblastoma cells; sodium butyrate increases the cAMP level in both mouse[81] and human neuroblastoma cells,[191] but 5-BrdU increase it only in mouse cells.[81] Both sodium butyrate and 5-BrdU do not inhibit cAMP phosphodiesterase activity; on the other hand, sodium butyrate-treated cells have elevated the level of adenylate cyclase activity.[48] Therefore, these agents increase intracellular level of cAMP by stimulating the adenylate cyclase activity. The increase in enzyme activity probably results from the extensive membrane changes observed after treatment of cells with these agents, since the effect on cAMP is not seen immediately after treatment. Serum-free medium is known to increase only some differentiated functions,[1,24] but it does increase cAMP level by twofold.[81] The mechanism of this effect is not known.

*Intracellular Level of cGMP*

The intracellular level of cyclic GMP in mouse neuroblastoma is 1% that of cAMP (Table 3-19). PGE₁ and Ro 20-1724 increase the

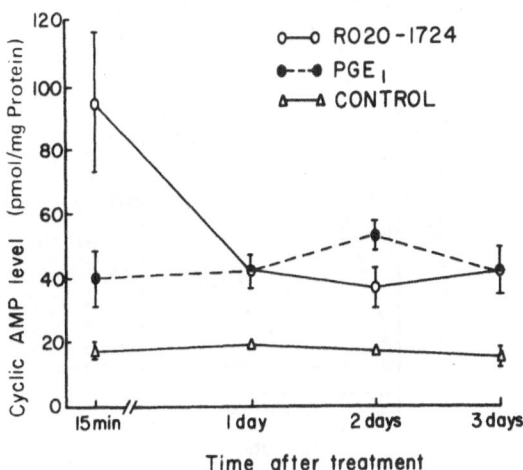

**Figure 3-15.** Changes in the cAMP level during morphological differentiation. Cells (0.5 × 10⁶) of clone NBA₂₍₁₎ were plated in 75-cm² Falcon plastic flasks, and PGE₁ (10 μg/ml) and Ro 20-1724 (200 μg/ml) were added separately 24 hr later. The control cultures received an equivalent volume of alcohol. The drug and medium were changed 2 days after treatment and the level of cAMP was determined 15 min, 1, 2, and 3 days after treatment. Each value represents an average of six to eight samples. The bar at each point is standard deviation. (From Prasad and Kumar.[19])

cAMP level without changing the cGMP level.[192] These measurements were made at 15 min after incubation. However, PGE₁ has been reported to increase cGMP level within 1 min and returned to normal level by 10 min (Figure 3-16). Thus, PGE₁ elicits a transitory, 30-sec increase in cGMP accumulation and a prolonged increase in cAMP accumulation.[193] The difference in the duration of cGMP and cAMP accumulations may reflect the activities of different enzymes which catalyze the synthesis and perhaps the hydrolysis of the nucleotides, or alternatively, may indicate the presence of two species of PGE₁ receptors, one mediating an increase in cAMP concentration, the other, an increase in cGMP. Matsuzawa and Nirenberg[193] have favored the latter idea. ACh increases the cGMP level without affecting the cAMP level in neuroblastoma cells, whereas papaverine increases both cyclic nucleotide levels (Table 3-19). A complex response is observed when different pairs of receptors were activated simultaneously (Table 3-20). In the presence of adenosine, cAMP levels increase and cGMP levels decrease, whereas in the presence of carbamycholine, cGMP levels increase and cAMP levels decrease. Thus, under the above experimental conditions, the inverse relationship between cGMP and cAMP concentration in neuroblastoma cells (NIE-115) exists. However, a novel phenomenon was observed. For example, in the presence of PGE₁, the levels of both cAMP and cGMP increase. The increase in cGMP elicited by carbamycholine was not inhibited by PGE₁, in spite of the fact that PGE₁ elicits a marked

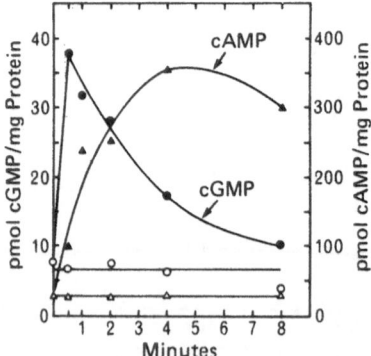

**Figure 3-16.** Effect of PGE₁ on intracellular levels of cGMP and cAMP in neuroblastoma NIE-115 cells. Cultures were grown for 8 days; the average cellular protein was 6.8 mg/100-mm dish. Closed and open circles correspond to cGMP levels with and without 10 μM PGE₁, respectively; closed and open triangles correspond to cAMP levels with and without 10 μM PGE₁, respectively. (From Matsuzawa and Nirenberg.[193])

**Table 3-19.** Effect of Various Agents on the Intracellular
Level of Cyclic Nucleotides in Neuroblastoma Cells[a]

| Treatment | Dose (μg/ml) | cAMP level (pmol/g protein) | cGMP level (pmol/g protein) |
|---|---|---|---|
| Control | | 5,000 ± 400 | 50 ± 7 |
| ACh | 100 | 5,300 ± 450 | 109 ± 12 |
| Ro 20-1724 | 200 | 29,000 ± 400 | 39 ± 6 |
| Papaverine | 25 | 32,000 ± 600 | 84 ± 9 |
| PGE₁ | 10 | 8,500 ± 500 | 41 ± 8 |

[a] From Prasad et al. [192]
[b] Values are mean ± S.D.

increase in cAMP levels. $PGE_1$- and carbamycholine-dependent increases in cGMP were fully additive, and in some experiments were greater than additive. In contrast, the elevated cAMP concentrations induced either by $PGE_1$ or by adenosine were not additive when both compounds were added simultaneously to cells. It was concluded[193] that activation of a species of muscarinic ACh receptor inhibits adenosine- and $PGE_1$-receptor-mediated reactions that elevate cAMP levels, whereas activation of adenosine receptors inhibits reactions mediated by muscarinic ACh receptors and $PGE_1$ receptors that elevate cGMP levels. In contrast to these reults, activation of $PGE_1$ receptors did not affect ACh-receptor-mediated reactions that affect cAMP or cGMP levels. Thus, reciprocal inhibition, unilateral inhibition, and additive and nonadditive responses were observed when different pairs of receptors were activated simultaneously. Relatively little is

**Table 3-20.** Summary of Receptor-Mediated Shifts
in cGMP and cAMP Levels in Neuroblastoma Cells[a]

| Addition | cGMP (%) | cAMP (%) |
|---|---|---|
| None | 100[b] | 100 |
| Adenosine | 40 | 500 |
| Carbamylcholine | 1500 | 70 |
| PGE₁ | 500 | 1250 |
| Adenosine + carbamylcholine | 650 | 250 |
| PGE₁ + carbamylcholine | 2050 | 1000 |
| PGE₁ + adenosine | 300 | 1100 |

[a] From Matsuzawa and Nirenberg. [193]
[b] 100% corresponds to 7.5 pmol cGMP/mg protein or 27.5 pmol cAMP/mg protein.

**Table 3-21.** Activities of cGMP and cAMP Phosphodiesterase in Mouse Neuroblastoma Cells in Culture[a]

| Enzymes | Low $K_m$ ($\mu$M) | $V_{max}$ of low $K_m$ (pmol/min/$\mu$g protein) | High $K_m$ ($\mu$M) | $V_{max}$ of high $K_m$ (pmol/min/$\mu$g protein) |
|---|---|---|---|---|
| cGMP phosphodiesterase (particulate) | 1.64 | 0.33 | 222.0 | 0.13 |
| cGMP phosphodiesterase (supernatant) | 3.6 | 1.07 | 44.0 | 1.46 |
| cAMP phosphodiesterase (particulate) | 4.1 | 0.90 | 100.0 | 1.9 |
| cAMP phosphodiesterase (supernatant) | 8.0 | 2.3 | 80.0 | 1.75 |

[a] The values of $K_m$ and $V_{max}$ for phosphodiesterase activities were determined by least-square linear regression of the double-reciprocal plot. The range of cyclic nucleotide concentrations varied from 0.3 $\mu$M to 1000 $\mu$M. From Prasad et al.[194]

known at the molecular level about the mechanisms which couple one receptor with another.

The regulation of intracellular levels of cAMP and cGMP by neurotransmitters is complex and remains to be fully understood. During differentiation of neuroblastoma cells, only prolonged increases in cAMP concentration are observed, whereas no significant change in the cGMP level is observed.[192]

*Regulation of Cyclic Nucleotide Phosphodiesterase Activity*

Phosphodiesterase activity in both particulate and supernatant fractions of control neuroblastoma cells (NBP$_2$ mouse, IMR-32) exists that hydrolyzes cyclic nucleotide with an apparent $K_m$ of 2–8 $\mu$M and with an apparent $K_m$ of 44–222 $\mu$M (Table 3-21). cAMP phosphodiesterase and cGMP phosphodiesterase are distinct enzymes and in part differently regulated.[194] The following evidence supports this: (1) Ro 20-1724 inhibits cAMP phosphodiesterase activity without changing the cGMP phosphodiesterase activity in homogenates of neuroblastoma cells; (2) cAMP phosphodiesterase activity increases in cAMP-induced differentiated neuroblastoma cells (Figure 3-17), but cGMP phosphodiesterase does not[146]; and (3) the specific activity of

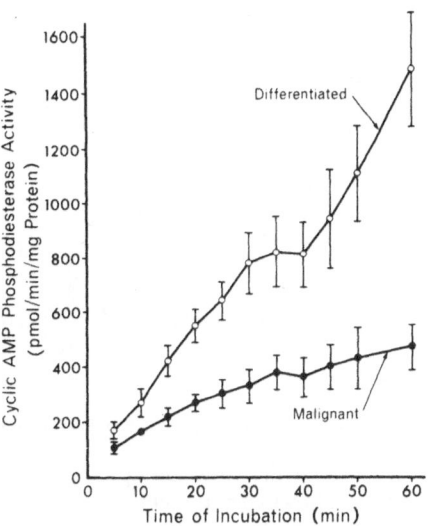

**Figure 3-17.** Effect of time of incubation on phosphodiesterase activity in homogenates of malignant and differentiated mouse neuroblastoma cells. Ro 20-1724 was used to induce differentiation. Enzyme activity was assayed in the presence of 30 mM magnesium. Each value represents an average of six samples. The bar at each point is standard deviation. (From Kumar *et al.*[198])

cGMP phosphodiesterase increases at protein concentrations of 1–4 $\mu$g/60 $\mu$l, reaches a plateau up to a concentration of 10 $\mu$g/60 $\mu$l, and then decreases,[194] whereas the specific activity of cAMP phosphodiesterase decreases as a function of protein concentration and a maximum inhibition in enzyme activity was achieved at 10 $\mu$g/60 $\mu$l protein concentration (Figure 3-18). Neuroblastoma cells appear to have endogenous inhibitor of cAMP phosphodiesterase. The following data support this: (1) The inhibition in enzyme activity occurs at protein concentrations which lie on the linear portion of the curve (Figure 3-19) when the total activity was plotted as a function of protein concentration. (2) At high protein concentrations the low $K_m$ values for cAMP phosphodiesterase increased (Table 3-22). (3) When the homogenate containing 10 $\mu$g protein/10 $\mu$l was boiled and then added (10 $\mu$g protein) to the reaction mixture containing 3 $\mu$g protein, the cAMP phosphodiesterase activity was not inhibited (Table 3-23). (4) When RNase- or DNase-treated homogenate containing 10 $\mu$g of protein was added to reaction mixture containing 3 $\mu$g protein, the cAMP phosphodiesterase activity was inhibited. The results suggest that the inhibitor substances are heat-sensitive but resistant to nuclease action. Although the presence of endogenous activator of cAMP

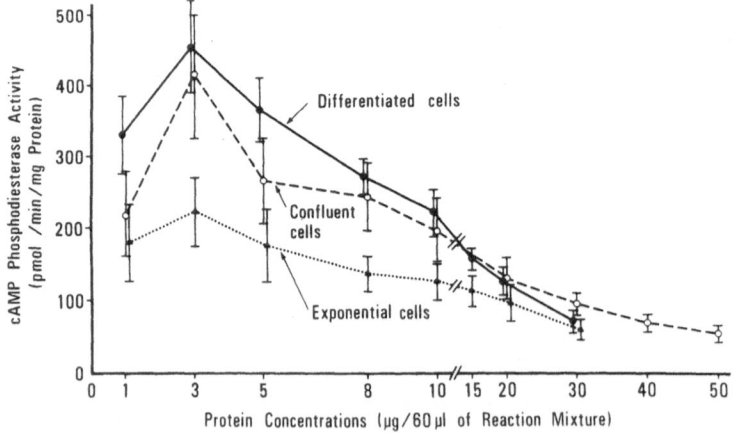

**Figure 3-18.** The specific activity of cAMP phosphodiesterase in homogenates of mouse neuroblastoma cells as a function of protein concentrations ($\mu$g/60 $\mu$l of reaction mixture). The reaction mixture (60 $\mu$l) contains 11,000 CPM; 3.33 $\mu$M or [³H]-cAMP. Cells (0.5 × 10⁶) were plated in 75-cm² Falcon flasks. Cells 2 days after plating were considered in exponential phase of growth, whereas cells 5 days after plating were considered in confluent phase of growth. Ro 20-1724 was used to induce differentiation. Each value represents an average of 10–12 samples. The bar at each point is standard deviation. (From Sinha and Prasad.[195])

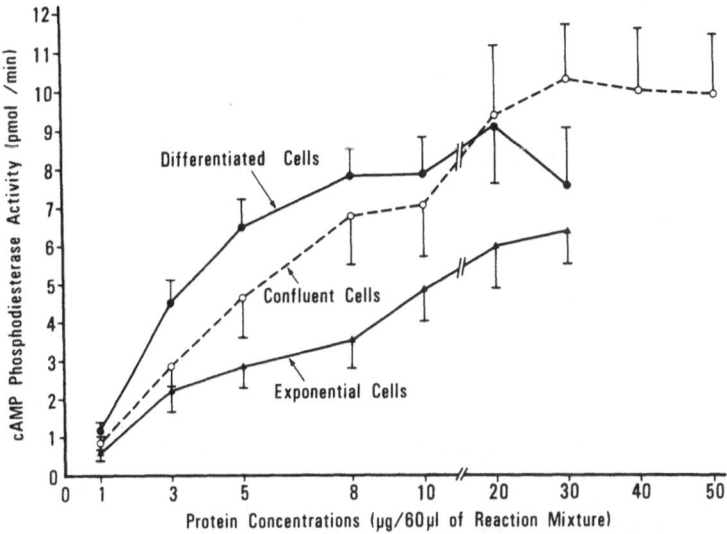

**Figure 3-19.** The total activity of cAMP phosphodiesterase as a function of protein concentrations ($\mu$g/60 $\mu$l of reaction mixture). The reaction misture (60 $\mu$l) contains 11,000 CPM; 3.33 $\mu$M of [³H]-cAMP. Cells (0.5 × 10⁶) were plated in 75-cm² Falcon flasks. Cells 2 days after plating were considered in confluent phase of growth. Ro 20-1724 was used to induce differentiation. Each value represents an average of 10–12 samples. The bar at each point is standard deviation. (From Sinha and Prasad.[195])

phosphodiesterase in mammalian brain has been demonstrated[196]; this is the first demonstration of endogenous inhibitor of cAMP phosphodiesterase in neuroblastoma cells. Whether such a phenomenon exists in nerve cells cannot be ascertained at this time.

**Table 3-22.** Effect of Protein Concentration on the $K_m$ Value of cAMP Phosphodiesterase Activity in Neuroblastoma Cells in Culture[a]

| $\mu$g protein/60 $\mu$l reaction mixture | $K_m$ ($\mu$M) | $V_{max}$ (pmol/min/$\mu$g protein) |
|---|---|---|
| 10.3 | 11.07 ± 5.1[b] | 0.73 ± 0.23 |
| 3.0 | 5.31 ± 2.2 | 0.73 ± 0.22 |
| 1.0 | 3.83 ± 1.38 | 1.02 ± 0.17 |

[a] Each experiment was repeated four times. The $K_m$ and $V_{max}$ values were determined by nonlinear least-square fitting of the data directly to the Michaelis-Menten equation using Marquardt's procedure.[23] The range of cAMP concentrations varied from 0.03 $\mu$M to 18 $\mu$M. From Prasad et al.[194]

Table 3-23.   Effect of Various Treatments on cAMP
Phosphodiesterase Activity in Mouse Neuroblastoma Cells[a]

| Treatment | cAMP phosphodiesterase activity (pmol/min/mg protein) |
|---|---|
| Control homogenate (3 μg protein) | 185 ± 13[b] |
| Control + untreated homogenate (10 μg protein) | 86 ± 10 |
| Control + boiled homogenate (10 μg protein) | 180 ± 8 |
| Control + DNase-treated homogenate (10 μg protein) | 79 ± 8 |
| Control + RNase-treated homogenate (10 μg protein) | 88 ± 7 |

[a] The reaction mixture (60 μl) contains 3 μg protein plus 10 μg untreated or treated protein. Each value represents an average of eight samples. From Prasad et al.[194]
[b] S.E.

## Changes in Cyclic Nucleotide Phosphodiesterase Activity during Differentiation

The activity of cAMP phosphodiesterase markedly increases in cAMP-induced differentiated cells,[197] whereas cGMP phosphodiesterase does not change.[195] Other agents, such as 6-TG, X irradiation, and 5'-AMP, which are known to increase some differentiated functions, fail to increase cAMP phosphodiesterase activity[197]; therefore, it has been suggested that cAMP may regulate the level of its own phosphodiesterase activity. The increase in cAMP phosphodiesterase activity occurs as early as 2 hr after dibutyryl cAMP treatment; the increase in enzyme activity after treatment with $PGE_1$ and Ro 20-1724 requires longer time, but by day 3 the values reached the same level irrespective of difference in treatment. The value of $K_m$ slightly decreases (Table 3-24), whereas the value of $V_{max}$ slightly increases (Table 3-25) in cAMP-induced differentiated cells. The pH optimum and the magnesium and manganese requirement for cAMP phosphodiesterase activity in control and differentiated cells were similar; however, the effect of certain agents, when assayed in the homogenate in the absence of exogenous magnesium, was different (Table 3-26). Calcium, zinc, copper, mercury, EDTA, and imidazole completely inhibited cAMP phosphosphodiesterase activity in control

**Table 3-24.** $K_m$ Values for Phosphodiesterase Activity in Neuroblastoma Cells in Culture[a]

| Neuroblastoma clones | Concentration ($\mu$M) | | | |
|---|---|---|---|---|
| | Low $K_m{}^b$ | Low $K_m{}^c$ | High $K_m{}^b$ | High $K_m{}^c$ |
| Mouse malignant (NBP$_2$) | 4.1 ± 0.7[d] | 12.5 ± 3.8 | 106 ± 53 | 259 ± 112 |
| Mouse differentiated (NBP$_2$) | 2.1 ± 0.4 | 7.0 ± 2.1 | 66 ± 27 | 108 ± 43 |
| Mouse malignant [NBE⁻(A)] | 2.9 | | 93 | |
| Mouse malignant (NBA$_{2(1)}$) | 2.4 | | 94 | |
| Human malignant (IMR-32) | 9.1 ± 1 | 20.4 ± 3.0 | 105 ± 34 | 169 ± 33 |

[a] Each value represents an average of five separate determinations, but the values of clones NBE⁻(A) and NBA$_{2(1)}$ represent only one determination. From Prasad and Kumar.[191]
[b] The value was determined by least-square linear regression of the double-reciprocal plot.
[c] The value was determined by nonlinear least-squares fitting of the data directly to the Michaelis-Menten equation using the procedure of Marquardt.[10]
[d] Mean ± S.E.

**Table 3-25.** $V_{max}$ Values for Phosphodiesterase Activity in Neuroblastoma Cells in Culture[a]

| Neuroblastoma clone | $V_{max}$ (pmol/min/$\mu$g protein) | | $V_{max}$ (pmol/min/$\mu$g protein) | |
|---|---|---|---|---|
| | Low $K_m{}^b$ | Low $K_m{}^c$ | High $K_m{}^b$ | High $K_m{}^c$ |
| Mouse malignant (NBP$_2$) | 0.07 ± 0.14[d] | 1.18 ± 0.27 | 2.86 ± 1.02 | 3.34 ± 1.4 |
| Mouse differentiated (NBP$_2$) | 0.66 ± 0.14 | 1.69 ± 0.24 | 5.11 ± 1.5 | 5.70 ± 2.4 |
| Mouse malignant [NBE⁻(A)] | 0.57 | | 2.58 | |
| Mouse malignant (NBA$_{2(1)}$) | 0.40 | | 2.30 | |
| Human malignant (IMR-32) | 1.42 ± 0.17 | 2.95 ± 0.83 | 6.1 ± 1.1 | 7.91 ± 1.1 |

[a] Each value represents an average of five separate determinations, but the values of clones NBE⁻(A) and NBA$_{2(1)}$ represent only one determination. From Prasad and Kumar.[191]
[b] The value was determined by least-square linear regression of the double-reciprocal plot.
[c] The value was determined by nonlinear least-squares fitting of the data directly to the Michaelis-Menten equation using the procedure of Marquardt.[10]
[d] Mean ± S.E.

**Table 3-26.** Effect of Ions, EDTA, and Imidazole on Phosphodiesterase Activity in Homogenates of Neuroblastoma Cells[a]

| | Basal activity | Ca | Zn | Fe | Cu | Hg | EDTA | Imidazole |
|---|---|---|---|---|---|---|---|---|
| Malignant | $34 \pm 10^b$ | $2.5 \pm 3.4$ | $1.2 \pm 2.7$ | $3.2 \pm 6$ | 0 | $6 \pm 7$ | 0 | $3.2 \pm 3.8$ |
| Differentiated | $65 \pm 19$ | $36 \pm 17$ | $30 \pm 10$ | $21 \pm 7$ | $24 \pm 5$ | $41 \pm 9$ | $2 \pm 3.3$ | $44 \pm 16$ |

[a] Ro 20-1724 was used to induce differentiation. The phosphodiesterase activity was assayed in the absence of exogenous magnesium. From Prasad and Kumar.[191]
[b] Mean ± S.D.

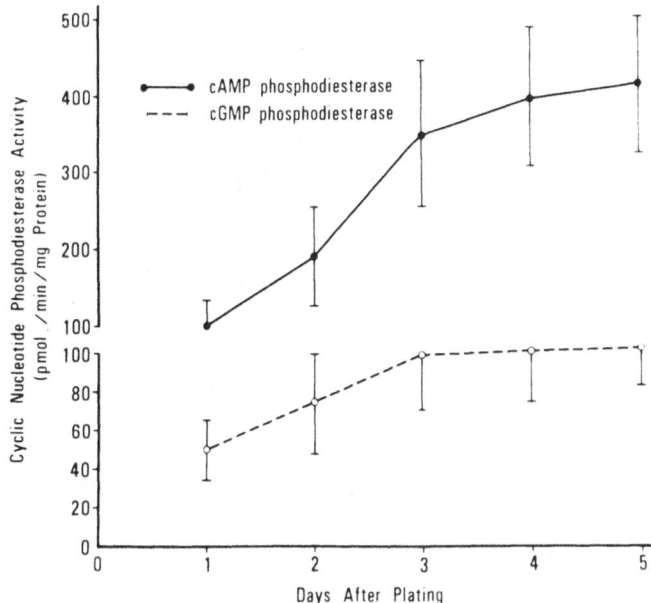

**Figure 3-20.** Cyclic nucleotide phosphodiesterase activity in homogenates of mouse neuroblastoma cells as a function of time after plating. The reaction mixture (60 μl) contains 3 μg protein, and 11,000 CPM (3.3 μM of [³H]-cAMP) or 10,000 CPM (3.3 μM of [³H]-cGMP). Each value represents an average of 12–15 samples. The bar at each point is standard deviation. (From Sinha and Prasad.[195])

cells, whereas the above agents, except EDTA, only partially inhibited enzyme activity in differentiated cells.[198]

The cAMP phosphodiesterase also increases during growth (Figure 3-20), but the intracellular level of cAMP does not change during growth. Hence, there are at least two possible mechanisms of regulation of cAMP phosphodiesterase—one involving change in cAMP; the other, growth. The increase in enzyme activity induced by cAMP is blocked by cycloheximide but not by actinomycin D,[197] whereas the increase in enzyme activity induced during growth is inhibited by both cycloheximide and actinomycin D.[195] The cGMP phosphodiesterase activity also increases slightly during the growth period.[195]

## Changes in the Cyclic Nucleotide Binding Proteins and cAMP-dependent Phosphorylation Activity during Differentiation

The current evidence is that the binding proteins constitute the regulatory subunits of cAMP-dependent protein kinase molecules.[199,202] The association of regulatory and catalytic subunits

makes the kinase molecules inactive. The binding of cAMP with the
regulatory subunits releases the catalytic subunits of kinase, which
then phosphorylate various proteins that are essential for the expression of cAMP effect on mammalian cells. The fact that cAMP-induced
differentiated cells not only have elevated levels of cAMP but also
have an increased level of cAMP phosphodiesterase activity suggests
that the cells must develop a mechanism of protecting the formed
cAMP from enzymatic hydrolysis. Indeed, the binding of cAMP with
soluble proteins increases[203] by about twofold (Figure 3-21). The protein-bound cAMP is protected from enzymatic hydrolysis. The binding of cAMP with protein of pellet (110,000$g$) and homogenates also
increased (Table 3-27). However, the extent of binding with soluble
proteins was much higher than that with pellet proteins. When the
soluble proteins from control and cAMP-induced differentiated cells,
containing bound and free [³H]-cAMP, were passed through a column of sephadex G-25 to resolve bound and free cAMP, the relative
amount of protein-bound [³H]-cAMP in differentiated cells was
greater than that in control cells, and the amount of free [³H]-cAMP was
correspondingly less (Figure 3-22). When the soluble proteins were
subjected to polyacrylamide gel electrophoresis, there were two binding peaks in both control and differentiated cells, but the extent of

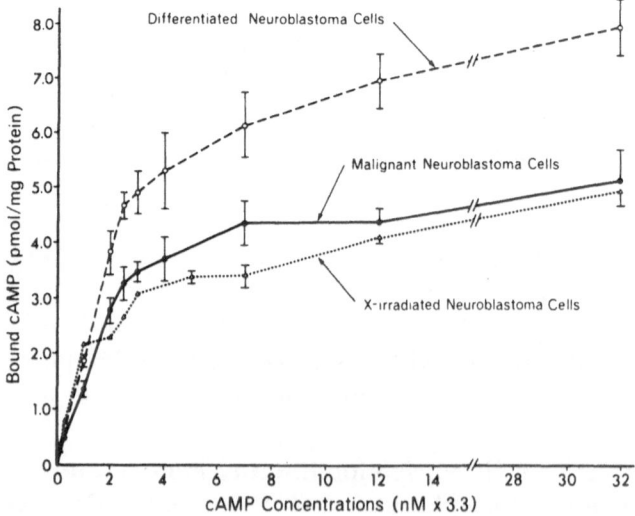

**Figure 3-21.** Binding of cAMP with soluble proteins of neuroblastoma cells as a function of cAMP concentrations. Ro 20-1724, and X irradiation (600 rads) were used to
induce differentiation. The incubation mixture contains 100 $\mu$g proteins. Each value
represents an average of six samples. The bar at each point is standard deviation. (From
Prasad *et al.*[204])

**Table 3-27.**  Binding of cAMP with Proteins of Neuroblastoma Cells[a]

| Treatment | Bound cAMP (pmol/mg protein) |
|---|---|
| Control Cells | |
| Soluble proteins | 4.20 ± 0.50[b] |
| Pellet proteins | 0.55 ± 0.04 |
| Total homogenate proteins | 2.38 ± 0.13 |
| Differentiated Cells | |
| Soluble proteins | 7.20 ± 0.60 |
| Pellet proteins | 1.33 ± 0.06 |
| Total homogenate proteins | 4.10 ± 0.12 |

[a] The cells were treated with Ro 20-1724 for 3 days. The binding assays were performed with soluble (100,000$g$, 90 min), pellet (100,000$g$, 90 min), and total homogenate proteins. The incubation mixture contained 100 $\mu$g proteins and 39.6 nM [³H]-cAMP. Each value represents an average of eight samples. From Prasad et al.[204]
[b] Mean ± S.D.

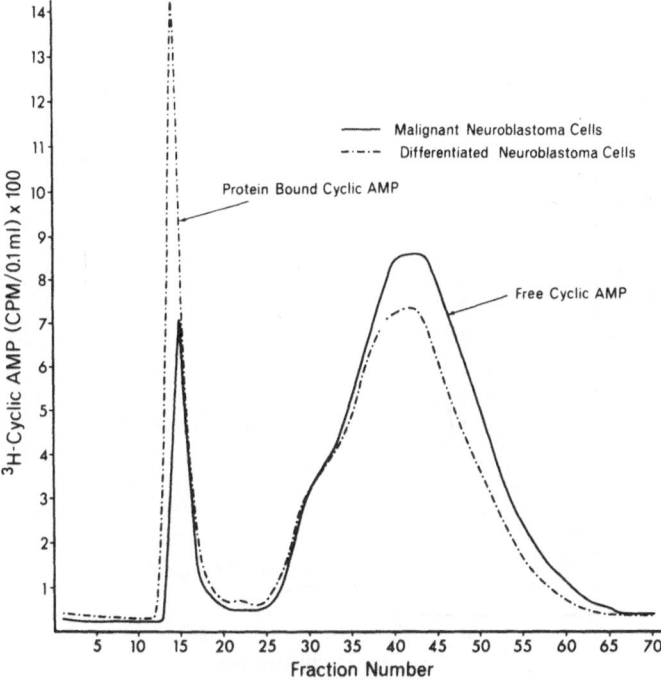

**Figure 3-22.** Gel filtration of protein-bound [³H]-cAMP through a sephadex G-25 column. Ro 20-1724 was used to induce differentiation. The soluble proteins (100,000$g$, supernatant) of malignant and differentiated neuroblastoma cells were incubated in the presence of [³H]-cAMP for 1 hr. The experiments were repeated twice. (Prasad et al.[204])

binding at each peak was relatively high in differentiated cells (Figure 3-23). The increase in cAMP binding occurred as early as 24 hr after PGE$_1$ treatment or 48 hr after Ro 20-1724 treatment, but no such increase in binding occurred in serum-free-medium-treated cells (Figure 3-24). The increase in binding was completely blocked by cycloheximide but not by actinomycin D (Table 3-28). The binding was heat labile and sensitive to protease action.

It has been suggested that cAMP may regulate the level of its own binding protein.[204] For example, PGE$_1$, 5-BrdU, and Ro 20-1724, which increase the intracellular level of cAMP,[1] elevate the level of cAMP binding proteins. However, X irradiation, 6-TG and 5'-AMP,

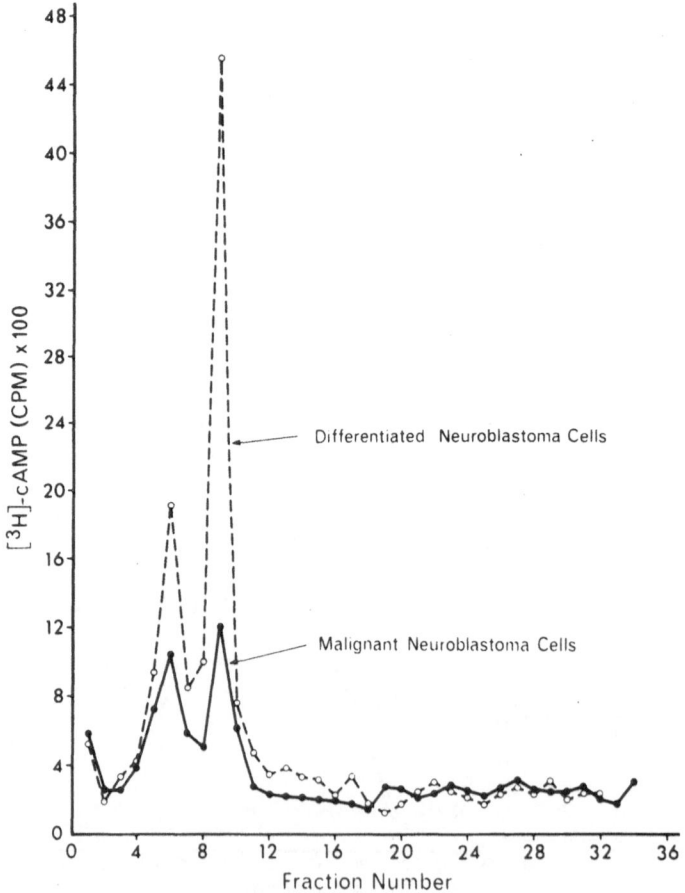

**Figure 3-23.** Binding of [$^3$H]-cAMP with protein fractions obtained after polyacrylamide gel electrophoresis. The soluble proteins of malignant and differentiated neuroblastoma cells were applied on the gel. Ro 20-1724 was used to induce differentiation. The experiments were repeated twice. (From Prasad *et al.*[203])

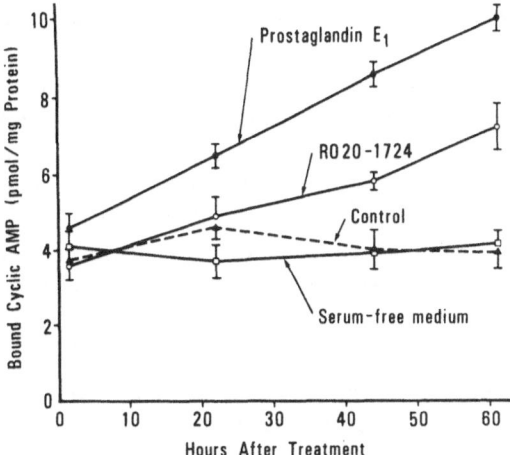

**Figure 3-24.** Binding of cAMP with soluble proteins of neuroblastoma cells as a function of treatment time. $PGE_1$, Ro 20-1724, and serum-free medium were added separately to culture, and the binding assay was performed 3 hr, 1 day, 2 days, and 3 days after treatment. The incubation mixture contains 100 $\mu$g proteins. Each value represents an average of six samples. The bar at each point is standard deviation. (From Prasad et al.[204])

which do not increase the cAMP level,[1] do not change the amount of binding proteins. cAMP may require an additional factor for increasing the levels of binding proteins in neuroblastoma cells because serum-free medium and sodium butyrate, which increase the cAMP level by about twofold, fail to increase the level of binding proteins.

**Table 3-28.** Effect of Metabolic Inhibitors on the Binding Capacity of cAMP with Soluble Proteins of Neuroblastoma Cells[a]

| Treatment | Bound cAMP (pmol/mg protein) |
|---|---|
| Control | $4.2 \pm 0.5$ |
| Control + cycloheximide | $3.3 \pm 0.2$ |
| Control + actinomycin D | $5.1 \pm 0.2$ |
| $PGE_1$ | $6.5 \pm 0.3$ |
| $PGE_1$ + cycloheximide | $3.5 \pm 0.3$ |
| $PGE_1$ + actinomycin D | $7.4 \pm 0.4$ |

[a] Control cells were treated with cycloheximide (5 $\mu$g/ml) or actinomycin D (5 $\mu$g/ml) for 12 hr before the binding assay was performed. Cells were treated with $PGE_1$ for a total of 24 hr after the addition of $PGE_1$, and cells were further incubated for an additional 12 hr. Each value represents an average of eight samples. From Prasad et al.[204]
[b] Mean $\pm$ S.D.

This second factor is not activated by cAMP when cells are treated with serum-free medium or sodium butyrate. An elevation in level of cAMP failed to increase the level of cAMP binding proteins in rat glioma and mouse L cells, indicating that either the levels of cAMP binding proteins in neuroblastoma cells and nonneural tumor cells are differently regulated, or that the second factor essential for the action of cAMP is missing in nonneural cells.[204]

It has been found[204] that neither the inhibition of cell division nor the increase in total protein is sufficient to increase the levels of cAMP binding proteins in mouse neuroblastoma cells. For example, 5'-AMP, X irradiation, and 6-TG inhibit cell division, but they do not increase the level of binding proteins. The amount of total protein in X-irradiated- and 6-TG-treated neuroblastoma cells increases (about threefold) to an extent similar to that observed in cAMP-induced differentiated neuroblastoma cells[18]; however, the levels of binding in these cells do not change.

cAMP and cGMP bind with the same proteins (Figure 3-25), but

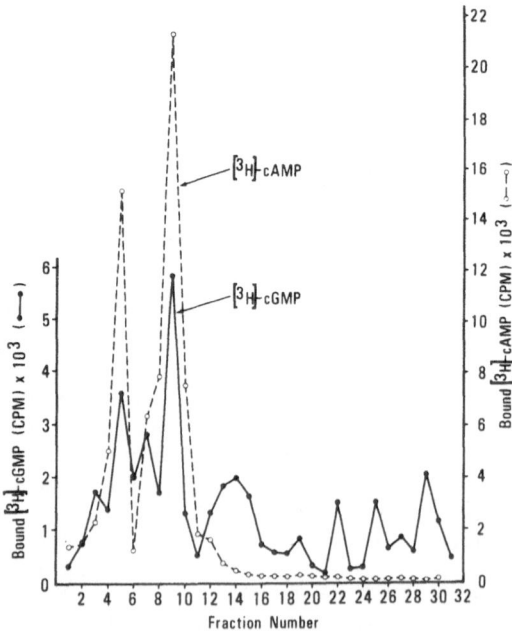

**Figure 3-25.** Binding of [³H]-cGMP and [³H]-cAMP with protein fractions obtained after polyacrylamide gel electrophoresis. The soluble proteins (1.5 mg) of malignant neuroblastoma cells were applied on the gel. The reaction mixture contained 7.7 nM [³H]-cAMP (specific activity 27 Ci/mmol) or 61 nM [³H]-cGMP (specific activity 21 Ci/mmol). The counting efficiency of liquid scintillation counter was 60%. The experiments were repeated twice. (From Prasad *et al.*[204])

the binding affinity of cAMP is 10 times more than that of cGMP. This is substantiated by the following observations: (1) The concentration of cGMP required to saturate the soluble protein is about 330 nM, whereas the concentration of cAMP required to saturate the soluble protein is about 33 nM; (2) the addition of 33 nM of nonradioactive cAMP to the reaction mixture reduces the binding of both [³H]-cAMP and [³H]-cGMP by 50%, whereas a concentration of 330 nM of non-radioactive cGMP is needed to produce a similar inhibition in binding; and (3) the pattern of binding of [³H]-cAMP and [³H]-cGMP with protein fractions (obtained after polyacrylamide gel electrophoresis) is similar. The intracellular level of cyclic GMP is 100 times less than that of cAMP. The low affinity of binding protein for cGMP may in part account for the low level of cGMP in neuroblastoma cells.

The significance of cAMP binding protein is not fully understood. In addition to being a regulatory subunit of cAMP-dependent protein kinase, it may provide one of the important intracellular mechanisms of maintaining high cAMP level during differentiation of neuroblastoma cells. The coexistence of increased cAMP level and increased cAMP phosphodiesterase activity is possible only in the presence of enhanced levels of cAMP binding proteins. This may also occur during development of normal brain cells, because the level of cAMP and the activity of cAMP phosphodiesterase increase in mature brain.[206-208] The increase in binding protein occurs in the dividing neuroblastoma cells,[204-205] but it may attain a maximal level at the time of inhibition of cell division. The effect of cAMP on nonneural tumor cells in culture is reversible. This may be due in part to the fact that cAMP fails to increase the cAMP binding proteins in nonneural tumor cells.[204]

If our present findings are applicable to *in vivo* conditions, the rationale for developing new therapeutic approaches for the treatment of neuroblastoma and nonneural tumors may be different. For example, an elevation of the intracellular level of cAMP may be of therapeutic value for neuroblastoma tumors, whereas an elevation of the intracellular level of cAMP binding proteins may be of therapeutic value for nonneural tumors.

Although the level of cAMP binding proteins increases in differentiated neuroblastoma cells, the activity of cAMP-dependent protein kinase of differentiated cells, when assayed in the presence of exogenous substrate ($H_1$-histone), did not change.[209] Since the activity of protein kinase towards histone is a poor measure of its phosphorylative activity towards endogenous natural substrate,[210] the phosphorylation activity of isolated protein bands on SDS gel electrophoresis was investigated.[79] The cytosol fraction revealed over 40

protein bands and no observed difference in staining patterns between control and Ro 20-1724-induced differentiated cells. Sixteen of
these proteins showed a detectable incorporation of [$^{32}$P]phosphate.
These phosphoprotein bands were designated according to their apparent molecular weight. Five major phosphorylated protein components (bands 55, 59, 72, 88, and 97) migrated in the region of 50,000–
100,000 daltons, and 11 bands that incorporated considerably less
labeled phosphate were distributed above and below this region. Of
the five major bands, the phosphorylation of one band (band 97) was
not affected by exogenous added cAMP. Phosphorylation of this
band decreased by about twofold in differentiated cells. The phosphorylation of the remaining band was cAMP-dependent and
showed a marked increase in differentiated cells (Figure 3-26). The
percentage stimulation of phosphorylation caused by the addition of
exogenous cAMP was approximately the same in differentiated and
control cells. The addition of exogenous cGMP stimulated phosphorylation in a manner similar to that produced by cAMP. However,
a 10-fold-higher concentration of cGMP was required. This finding is
consistent with the observation that cAMP and cGMP bind to the
same proteins in neuroblastoma cells but the binding affinity of cGMP
is 10 times less than that of cAMP.[203]

The profiles of electrophoretically separated proteins of the
cytosol, the crude nuclear fraction, and the pellet fraction as revealed
by staining with Coomassie blue were clearly different (Figure 3-27).
Endogenous phosphorylation activity in the nuclear and pellet fractions was much greater than that in the cytosol. The particulate fraction showed 18 labeled phosphoprotein bands, which demonstrated a
varied response to the presence of added cAMP in the phosphorylation assay. The protein-staining patterns of the particulate fraction
from control and differentiated cells were similar, phosphorylation in
most of the phosphoprotein bands was somewhat decreased in the
differentiated cells, but the amount of cAMP bound to protein in this
fraction did not change in differentiated cells.

The nuclear fraction of control cells showed two phosphorylation
bands (one prominent band, 112; one faint band, 100), while the
nuclear fraction of differentiated cells showed two predominant
bands (bands 112 and 100). The phosphorylation of these bands was
independent of cAMP. Quantification of these bands by densitomet-
ric tracing of radioautogram indicated that the phosphorylation of
both bands in differentiated cells markedly increased (Figure 3-27).

The increase in cAMP-dependent phosphorylation activity in the
cytosol of differentiated cells correlated well with the increase in the

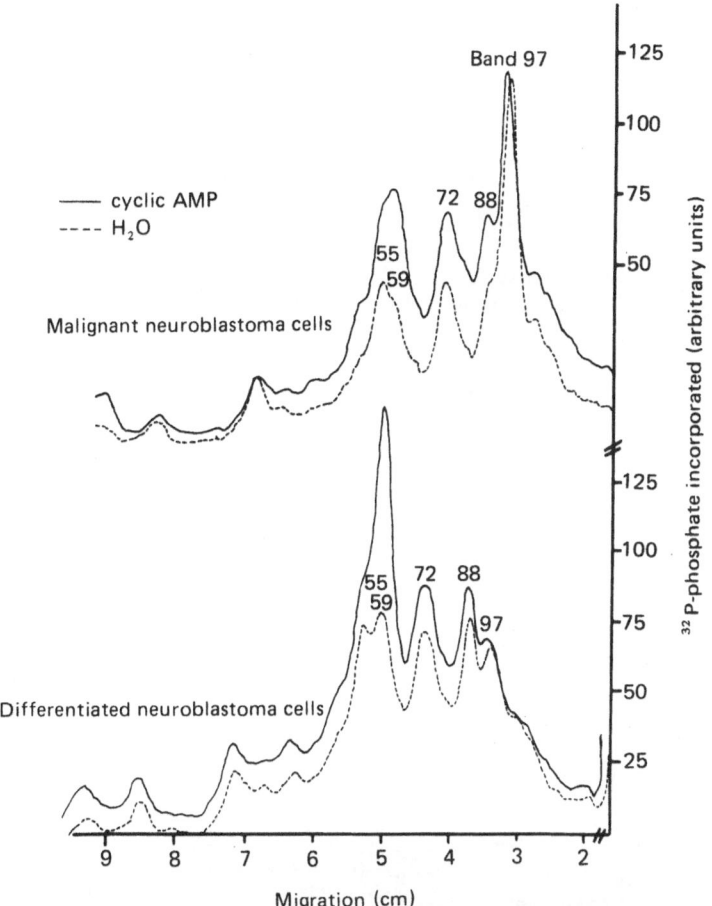

**Figure 3-26.** Autoradiogram tracing of phosphorylated proteins in the cytosol of malignant and cAMP-induced differentiated neuroblastoma cells in culture. It would be noted that the incorporation of $^{32}P$ into proteins from gamma [$^{32}P$]-ATP is a net result of the action of activities of protein kinase(s) and phosphoprotein phosphotase(s). Therefore, the term phosphorylation refers to changes in the amount of radioactive phosphate incorporation into specific bands under standard assay conditions, and does not refer to the capacity of specific enzymes. (From Ehrlich et al. [79])

binding of cAMP by protein localized in this fraction. A previous study[203] has shown that two peaks of cAMP binding proteins can be observed on polyacrylamide gels and that the extent of binding at each peak was markedly increased in differentiated cells. It is unknown whether the cAMP-binding proteins of neuroblastoma cells constitute only regulatory subunits of cAMP-dependent protein kinase or whether some of them can directly affect nuclear events

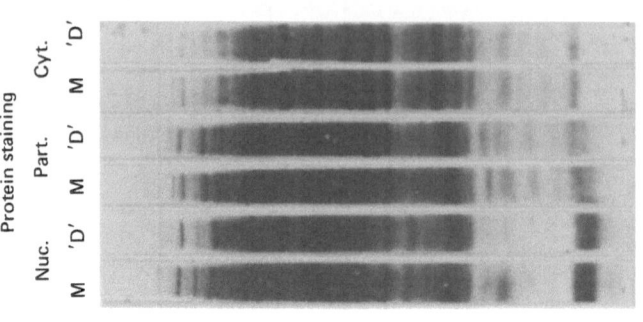

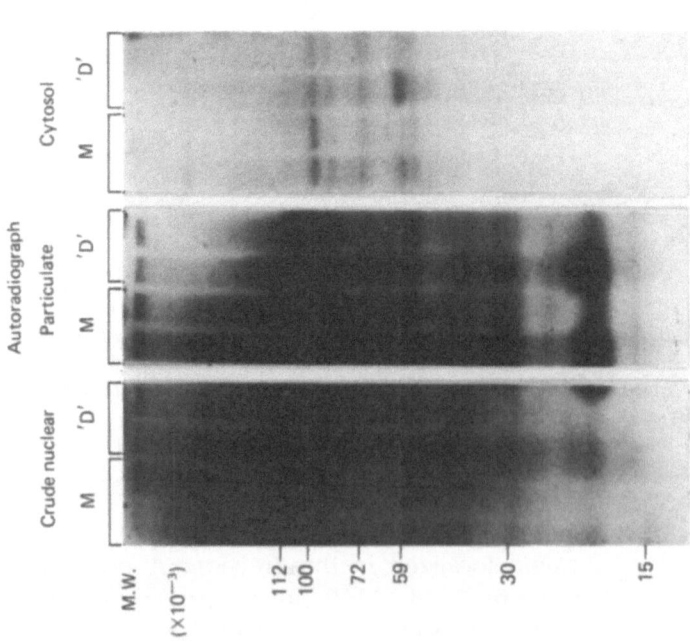

**Figure 3-27.** Autoradiogram of phosphorylated proteins of various subcellular fractions obtained from malignant and cAMP-induced differentiated neuroblastoma cells in culture. It should be noted that the incorporation of $^{32}$P into proteins from gamma [$^{32}$P]-ATP is a net result of the action of activities of protein kinase(s) and phosphoprotein phosphatase(s). Therefore, the term phosphorylation refers to changes in the amount of radioactive phosphate incorporation into specific bands under standard assay conditions, and does not refer to the capacity of specific enzymes. (From Ehrlich et al.[79])

during differentiation. Proteins that bind cAMP and regulatory sub-units of protein kinase are not identical molecular species.[6,7] It has been shown[211] that cAMP stimulates the assembly of nuclear microtubules and microfilaments in sympathetic neurons. The mechanism of this effect is unknown.

The assay of phosphorylation activity in neuroblastoma cells has been studied by three methods: incorporation of $^{32}$P in the intact cells, incorporation of $^{32}$P in isolated protein fractions of cells, and incorporation of $^{32}$P in isolated proteins using exogenous calf thymus histone fractions as substrate. A markedly different result can be obtained by these techniques. For example, using exogenous histone fractions in the cytosol of neuroblastoma cells, it has been shown that the addition of cAMP stimulates phosphorylation activity in the cytosol in the presence of $H_1$-histone and $H_2b$ histone but not in the presence of $H_2a$ histone. The extent of phosphorylation in the cytosol of differentiated cells decreased by twofold when histone $H_2b$ was used as a substrate (unpublished observation), but showed no significant change when $H_1$-histone was used as a substrate.[209] Thus, one may conclude that the cAMP-dependent phosphorylation decreases in the cytosol of differentiated cells. However, when the endogenous isolated protein fractions were used, the cAMP-dependent phosphorylation activity in the cytosol of differentiated cells markedly increased.[79] the cAMP-dependent phosphorylation activity could not be demonstrated in the nuclear fraction when the incorporation of $^{32}$P in the intact cells was studied after an increase in the intracellular level of cAMP,[78] or when the incorporation of $^{32}$P in the isolated protein fraction from the crude nuclear fraction was measured.[79] It is not known if there is no cAMP-dependent protein kinase in the nucleus or if the method is insensitive. From these data it is clear that the phosphorylation study must be performed both *in vitro* and *in vivo* before any conclusion with respect to its role in regulation of growth and differentiation can be evaluated. Thus, the exact mechanism of cAMP on the neuroblastoma cells remains unknown, but it somehow increases the expression of certain genes and suppresses the expression of others. The changes in the binding proteins as well as in cAMP-dependent and cAMP-independent phosphorylation are involved.

Based on data on mouse neuroblastoma cells, one can make a diagrammatic representation of possible events involved in the malignancy of nerve cells (Figure 3-28). The reverse change could have happened if the cells were to go to normal differentiation. The model[212] described below suggests that a mutation in the regulatory gene for cAMP phosphodiesterase with ⌐ ⌐ single and/or a group of

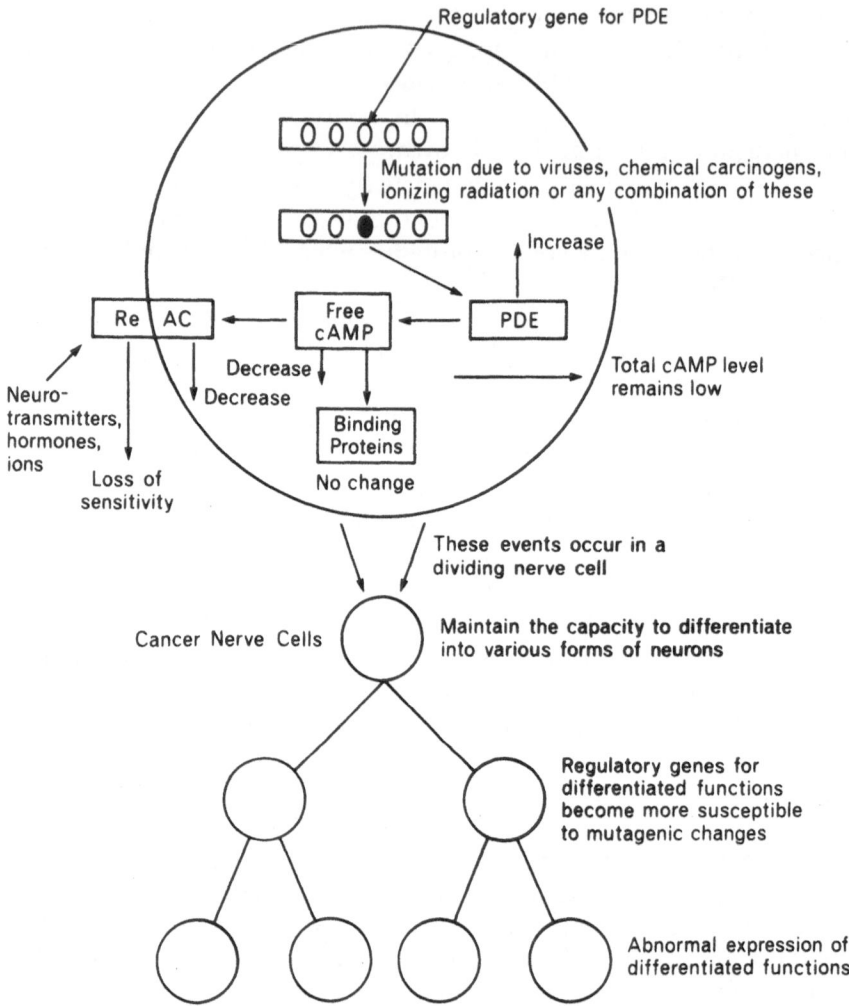

**Figure 3-28.** A diagrammatic model to explain the postulated mechanism for the development of cancer of nerve cells. Adenosine, 3',5'-cyclic monophosphate = cAMP; cAMP phosphodiesterase = PDE; adenylate cyclase = AC; receptor = Re. (From Prasad.[212])

dividing nerve cells may result from viruses, chemical carcinogens, ionizing radiation, or any combination of these agents, and this mutational change may increase phosphodiesterase activity in mutated cells. The high cAMP phosphodiesterase activity could lead to low levels of cAMP, which in turn could cause low levels of adenylate cyclase activity, and insensitivity of adenylate cyclase to neuro-transmitters. These changes might then prevent the expression of neuronal differentiated functions in the mutated nerve cell, causing it to become a cancer cell. The regulatory genes for other differentiated functions in the daughter cancer cells may consequently become more susceptible to mutagenic changes, which then may produce further molecular lesions. This may account for the fact that the neuroblastoma cells obtained from a tumor differ quantitatively and qualitatively from one another with respect to expression of cellular properties and sensitivity to different drugs. The mutated nerve cell appears to maintain the capacity to differentiate into various forms of nerve cells. This is supported by the fact the neuroblastoma contain four major types of nerve cell[3,80,213]: (1) adrenergic cells; (2) cholinergic cells; (3) sensorylike cells; and (4) serotonin cells. However, the differentiated functions of these nerve cells are not adequately expressed, and therefore they continue to divide. Whether or not the first cancer nerve cell will lead to the formation of a detectable neoplasm depends upon the host's immunological environment. If the host's immunological environment is normal, the mutated nerve cell may be rejected and no malignant lesion will ever appear. On the other hand, if the immunological environment of the host is abnormal, the malignant neoplasm may become detectable in a few months or years after the appearance of the first cancer cell. The following experimental data support the proposed hypothesis: (1) A raised intracellular level of cAMP in neuroblastoma cells irreversibly induces several differentiated functions characteristic of mature neurons[1]; (2) the inhibition of cAMP phosphodiesterase activity reduces the oncogenicity of differentiated cells[1]; (3) although DA- and NE-sensitive adenylate cyclases are demonstrable *in vitro*,[48] these agents do not increase the cAMP level until the phosphodiesterase activity is inhibited[134-187]; and (4) a relatively high phosphodiesterase activity in neuroblastoma cells is associated with a low activity of adenylate cyclase and a low level of cAMP.[188] Although the basal level of cAMP phosphodiesterase activity in differentiated mouse neuroblastoma cells increases,[197,198] the intracellular level of cAMP also increases.[19] The association of a high level of cAMP with the high level of cAMP phosphodiesterase activity may be due to the increased level of cAMP binding proteins.[203] The protein-bound cAMP may be relatively less susceptible to enzymatic hydrolysis.

Our evidence does not suggest that a defect in the adenylate cyclase regulation is one of the early lesions of the malignancy of nerve cells. For example, the adenylate cyclase activity in neuroblastoma homogenates can be stimulated by DA, NE, PGE₁, and ACh[1]; however, only $PGE_1$ increases the intracellular level of cAMP without the inhibition of AMP phosphodiesterase activity.[81] In one clone of neuroblastoma cells, even $PGE_1$ does not increase the cAMP level until the phosphodiesterase activity is inhibited.[187] Thus the rate-limiting factor in the accumulation of cAMP after treatment of cells with dopamine, NE, $PGE_1$, and adenosine is the high activity of phosphodiesterase.

The present model assumes that the malignant transformation occurs in a single diving nerve cell and/or in a group of dividing nerve cells. There is some indirect evidence to support this. For example, these tumors are generally detectable in children under five years of age. In addition, the neuroblastomas contain more than one type of nerve cell, indicating that the malignant transformation occurs in nerve cells which are not yet committed to form any one neural cell type. There is no evidence that the mature neurons ever undergo malignant transformation, although the supporting elements of adult nervous tissue do so.

## EFFECT OF VITAMIN C

Mouse neuroblastoma (Figure 3-29) cells are more sensitive to sodium ascorbate (vitamin C) than mouse fibroblasts (L cells) and rat glioma for the criterion of growth inhibition (due to cell death and reduction in cell division). Sodium L-ascorbate at nonlethal concentrations potentiates the effect of 5-fluorouracil (5-FU), X irradiation, bleomycin, Ro 20-1724, $PGE_1$, and sodium butyrate on NB cells,[216] but does not enhance the effect of vincristine, 6-TG, or CCNU except at higher drug doses.[215] On the other hand, the cytotoxic effect of DTIC is completely prevented by sodium ascorbate; however, the effect of methotrexate is only partially prevented by sodium L-ascorbate. If the present results could be applicable to human tumor *in vivo*, the addition of sodium L-ascorbate with certain chemotherapeutic agents may markedly increase their cytotoxic effect on tumor cells in a highly selective manner without affecting any further host's immune system. The addition of catalase (200 $\mu$g/ml) completely prevents the cytotoxic effect of high concentrations of sodium ascorbate, but it fails to prevent the potentiating effect of low concentrations of sodium L-ascorbate. Sodium L-ascorbate inhibits catalase activity *in vitro*.[216] Data indirectly indicate that catalase activity in neuroblastoma cells may be a predominant mode of destroying cellular $H_2O_2$. We there-

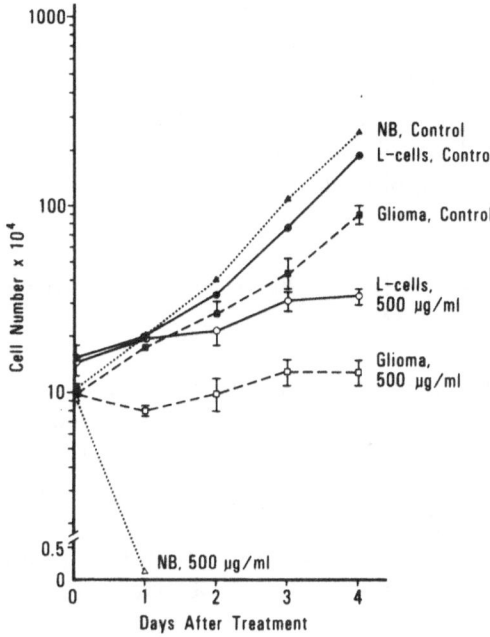

**Figure 3-29.** Effect of sodium ascorbate on mammalian cells in culture.

fore speculate that the tumor cells in which catalase is the predominant mode of destroying $H_2O_2$ would respond to the potentiating effect of sodium ascorbate in a manner observed for neuroblastoma cells in culture.

## CAN DIFFERENTIATED CELLS ACT AS A STRONG ANTIGEN?

Human neuroblastoma cells in culture possess tumor-specific transplantation antigens against which an immune reaction can be demonstrated with the colony inhibition assay.[217] We have obtained preliminary evidence indicating that the differentiated cells may act as a strong antigen. Uncloned cells treated with $PGE_1$ and Ro 20-1724 for four days do not produce tumors in all cases. Many of the mice which fail to develop tumors after subcutaneous injection of differentiated cells reject the subcutaneously administered malignant neuroblastoma cells $(0.25 \times 10^6)$.[188] The differentiated cells may retain the tumor-specific antigens and therefore may eventually reject themselves. The above phenomenon remains to be fully established. However, it has been observed clinically that the spontaneous regression of neuroblastoma tumors proceeds with the transformation to ganglioneuromas, which also eventually disappear. If these assumptions are correct, the agent which causes differentiation of neuroblastoma cells may be useful in the management of human neuroblastoma tumors in a highly selective manner.

## SUGGESTION OF A NEW APPROACH FOR THE TREATMENT OF NEUROBLASTOMAS

Based on the current data on neuroblastoma cells in culture, the previous suggestion for the therapy[134] of neuroblastoma tumor should be modified as follows: (1) Sodium ascorbate and sodium butyrate are given daily for the entire period of treatment. Sodium ascorbate is a relatively nontoxic compound. The oral administration of sodium butyrate (6–10 g/day) for four months produced no detectable toxic effects in a child with neuroblastoma (L. Furman, personal communication). (2) After 5 days of above treatment, cyclophosphamide, vincristine, adriamycin, and 5-fluorouracil are administered in a sequence, and in dosages and time intervals which are currently used. It is hoped that after the completion of the second phase, most of the tumor cells would be killed. (3) After the second phase, the differentiating agents such as cAMP-stimulating agents (inhibitors of phosphodiesterase like papaverine and Ro 20-1724, and a stimulator of adenylate cyclase like prostaglandin $E_1$) are continuously infused for at least 4 days. The administration of $PGE_1$ in the presence of an inhibitor of cAMP phosphodiesterase is suggested because the neuroblastoma tumor contains cells whose sensitivity to $PGE_1$ and phosphodiesterase inhibitor markedly varies. However, the combination of $PGE_1$ and phosphodiesterase inhibitor increases the intracellular level of cAMP in all clones.[187] It is hoped that the administration of differentiating agents would produce differentiation in many of the remaining cells. The differentiated cells may evoke an immune response which would kill many of the drug-resistant tumor cells. (4) After the administration of differentiating agents, the nonspecific immune stimulants such as BCG should be administered. It is hoped that a further stimulation of the host's immune system would reject the residual drug-resistant tumor cells. The phases 1, 3, and 4 should be repeated at least twice after the clinical free state has been achieved. Because of the carcinogenic effect of ionizing radiation, the use of radiation for the treatment of neuroblastoma is not recommended until an extreme emergency condition exists.

## CLINICAL TRIAL OF DIFFERENTIATING AGENTS

Dr. L. Helson of Memorial Hospital, New York, has been using combinations of cytotoxic drugs such as cyclophosphamide and vincristine in combination with differentiating agents such as papaverine and trifluoro-methyl-2-deoxyuridine in metastatic neuroblastoma patients. Although a marked regression of tumor was observed in all patients, and the conversion from neuroblastoma to ganglioneuroma was observed in cases where biopsies were taken and examined,[28] the response of the untreated patients appears to be the best. Dr. Helson has indicated (personal communication) that 17 out of 19 untreated patients remain free of disease between 12 and 24 months. Dr. Imashuku et al.[218] have shown that the levels of cAMP in ganglioneuromas were eightfold higher than those found in immature round cell neuroblastomas. These values were similar to those observed in adult sympathetic ganglia. Thus, there is a good correlation between higher levels of cAMP and higher levels of differentiation in neuroblastoma cells in vivo. It has also been shown[218] papaverine increases the intracellular levels of cAMP in immature round cell tumor in vivo, a phenomenon already demonstrated in vitro.[190]

## CONCLUSION

The following conclusions can be made. (1) cAMP appears to be one of the important factors in induction as well as in regulation of several differentiating functions in mammalian nerve cells. A low level of cAMP in dividing nerve cells may result from an increase in cAMP phosphodiesterase activity due to a mutation on the regulatory gene of this enzyme and may be responsible for the expression of malignancy and "abnormal differentiation." (2) The exact relationship between differentiation and malignancy remains to be clarified. However, it appears that no one individual differentiated function is linked with malignancy, which is not surprising, because the expression of many of these functions is independently regulated. However, when several of these differentiated functions express at maximal levels, the tumorigenicity of such cells is abolished. Thus, the expression of malignancy and abnormal differentiation appeared to be linked in neuroblastoma cells in culture. (3) The increased level of cAMP binding proteins provides one the important intracellular mechanisms of protecting the formed cAMP from enzymatic hydrolysis during differentiation of neuroblastoma cells. (4) The reduction in histone synthesis and in $H_1$-histone phosphorylation may be an important biological signal for the dividing neuroblasts to turn off cell division. If these events do not occur, that might be indicative of malignant change in dividing neuroblasts.

## REFERENCES

1. Prasad, K. N., Differentiation of neuroblastoma cells in culture, Biol. Rev. 50:129–165, 1975.
2. Wahn, H. L., Lightbody, L. E., and Tchen, T. T., Induction of neural differentiation in culture of amphibian undetermined presumptive epidermis by cyclic AMP derivatives, Science 188:366–369.
3. Amano, T., Richelson, E., and Nirenberg, M., Neurotransmitter synthesis by neuroblastoma clones, Proc. Natl. Acad. Sci. U.S.A. 69:258–263, 1972.
4. Biedler, J. L., Helson, L., and Spengler, B. A. Morphology and growth tumorigenicity, and cytogenetics of human neuroblastoma cells in continuous culture, Cancer Res. 33:2643–2652, 1973.
5. Schubert, D., Heinemann, S., Carlisle, W., Tarikas, H., Kimes, B., Patrick, J., Steinbach, J. H., Culp, W., and Brandt, B. L., Clonal cell lines from rat central nervous system, Nature (London) 249:224–227, 1974.
6. Chambaut, A. M., Leray, F., and Hanoune, J., Relationship between cyclic AMP dependent protein kinase(s) and cyclic AMP binding protein(s) in rat liver, FEBS Lett. 15:328–334, 1971.
7. Lee, P. C., and Jungmann, R. A., Ontogeny of cyclic AMP-dependent protein phosphokinase during hepatic development of the rat, Biochim. Biophys. Acta 399:265–276, 1975.
8. Prasad, N., Prasad, R., and Prasad, K. N., Electrophoretic patterns of glucose metabolizing enzymes and acid phosphatase in mouse and human neuroblastoma cells, Exp. Cell Res. 104:273–277, 1977.
9. Ciesielski-Treska, J., Mandel, P., Tholey, G., and Wurtz, B., Enzymatic activities modified during multiplication and differentiation of neuroblastoma cells, Nature (London), New Biol. 239:180–181, 1972.
10. Dawson, G., and Stoolmiller, A. C., Comparison of ganglioside composition of established mouse neuroblastoma cell strains grown in vivo and in tissue culture, J. Neurochem. 263:225–226, 1976.
11. Augusti-Tocco, G., and Sato, G., Establishment of functional clonal lines of neurons from mouse neuroblastoma, Proc. Natl. Acad. Sci. U.S.A. 64:311–315, 1969.
12. Waymire, J. C., Weiner, N., and Prasad, K. N., Regulation of tyrosine hydroxylase activity in cultured mouse neuroblastoma cells. Elevation induced by

analogs of adenosine 3', 5'-cyclic monophosphate, *Proc. Natl. Acad. Sci. U.S.A.* **69:**2241–2245, 1972.

13. Prasad, R., Prasad, N., and Prasad, K. N., Esterase, malate, and lactate dehydrogenase activity in murine neuroblastoma, *Science* **181:**450–451, 1973.

14. Tholey, G., Wurtz, B., Ciesielski-Treska, J., and Mandel, P., Lactate dehydrogenase in neuroblastoma clones, *J. Neurochem.* **23:**1083–1084, 1974.

15. Prasad, K. N., Differentiation and growth of neuroblastoma cells and serum types, *Trans. Am. Soc. Neurochem.* **87** (Abstr.), 1977.

16. Van Der, L. H., The "improperly" oriented pyramidal cell in the cerebral cortex and its possible bearing on the problem of neuronal growth and cell orientation, *Bull. Johns Hopkins Hosp.* **117:**228–250, 1965.

17. Bray, D., Branching patterns of individual sympathetic neurons in culture, *J. Cell Biol.* **56:**702–712, 1973.

18. Nelson, P., Ruffner, W., and Nirenberg, M., Neuronal tumor cell with excitable membranes grown *in vitro*, *Proc. Natl. Acad. Sci. U.S.A.* **64:**1004–1110, 1969.

19. Prasad, K. N., and Kumar, S., Cyclic AMP and the differentiation of neuroblastoma cells in culture, in: *Control of Proliferation in Animal Cells* (B. Clarkson and R. Baserga, eds.), pp. 581–594, Cold Spring Harbor Laboratory, Cold Spring Harbor, N.Y., 1974.

20. Prasad, K. N., and Hsie, A. W., Morphological differentiation of mouse neuroblastoma cells induced *in vitro* by dibutyryl adenosine 3':5' cyclic monophosphate, *Nature (London) New Biol.* **233:**141–142, 1971.

21. Furmanski, P., Silverman, D. J., and Lubin, M., Expression of differentiated functions in mouse neuroblastoma mediated by dibutyryl cyclic adenosine monophosphate, *Nature (London)* **233:**413–415, 1971.

22. Prasad, K. N., Morphological differentiation induced by prostaglandin in mouse neuroblastoma cells in culture, *Nature (London), New Biol.* **236:**49–52, 1972.

23. Prasad, K. N., and Sheppard, J. R. Inhibitors of cyclic nucleotide phosphodiesterase induce morphological differentiation of mouse neuroblastoma cell culture, *Exp. Cell. Res.* **73:**436–440, 1972.

24. Seeds, N. W., Gilman, A. G., Amano, T., and Nirenberg, M. W., Regulation of axon formation by clonal lines of a neural tumor, *Proc. Natl. Acad. Sci. U.S.A.* **66:**160–167, 1970.

25. Prasad, K. N., X-ray induced morphological differentiation of mouse neuroblastoma cells *in vitro*, *Nature (London)* **234:**471–474, 1971.

26. Schubert, D., and Jacob, F., 5-bromodeoxyuridine-induced differentiation of a neuroblastoma, *Proc. Natl. Acad. Sci. U.S.A.* **67:**247–254, 1970.

27. Prasad, K. N., Mandal, B., and Kumar, S., Human neuroblastoma cell culture: Effect of 5-bromodeoxyuridine on morphological differentiation and levels of neural enzymes, *Proc. Soc. Exp. Biol. Med.* **144:**38–42, 1973.

28. Helson, L., Management of disseminated neuroblastoma, *Ca* **25:**264–268, 1975.

29. Prasad, K. N., Differentiation of neuroblastoma cells induced in culture by 6-thioguanine, *Int. J. Cancer* **12:**631–635, 1973.

30. Kates, J. R., Winterton, R., and Schlesinger, K., Induction of acetylcholinesterase activity in mouse neuroblastoma tissue culture cells, *Nature (London)* **229:**345–346, 1971.

31. Byfield, J. E., and Karlsson, U., Inhibition of replication and differentiation in malignant mouse neuroblasts, *Cell Differ.* **2:**55–64, 1973.

32. Monard, D., Solomon, F., Rentsch, M., and Gysin, R., Glial-induced

morphological differentiation in neuroblastoma cells, *Proc. Natl. Acad. Sci. U.S.A.* **70**:1894–1897, 1973.

33. Reynolds, C. P., and Perez-Polo, J. R., Human neuroblastoma: Glial induced morphological differentiation, *Neurosci. Lett.* **1**:91–97, 1975.

34. Ross, J., Granett, S., and Rosenbaum, J. L., Differentiation of neuroblastoma cells in hypertonic medium, *J. Cell Biol.* **59**:291a, 1973.

35. Goldstein, M. N., Land, V., and Bradshaw, R., Stimulation of human neuroblastomas *in vitro* with nerve growth factor, *Proc. Am. Assoc. Cancer Res.* **3**:89, a, 1972.

36. Waris, T., Richard, L., and Waris, P., Differentiation of neuroblastoma cells induced by nerve growth factor *in vitro*, *Experientia* **29**:1128–1129, 1973.

37. Furmanski, P., and Lubin, M., Effects of dimethysulfoxide on expression of differentiated functions in mouse neuroblastoma, *J. Natl. Cancer Inst.* **48**:1355–1361, 1972.

38. Prasad, K. N., Effect of cytochalasin B and vinblastine on x-ray, dibutyryl cyclic AMP and prostaglandin-induced differentiation of mouse neuroblastoma cell culture, *Cytobios* **5**:265–271, 1972.

39. Hinkley, R. E., and Telser, A. G., The effect of halothane on cultured mouse neuroblastoma cells. Inhibition of morphological differentiation, *J. Cell. Biol.* **63**:531–540, 1974.

40. Sheppard, J. R., and Prasad, K. N., Cyclic AMP levels and the morphological differentiation of mouse neuroblastoma cells, *Life Sci.* **12**:431–439, 1973.

41. Miller, R. A., and Ruddle, F. H., Enucleated neuroblastoma cells form neurites when treated with dibutyryl cyclic AMP, *J. Cell Biol.* **63**:295–299, 1974.

42. Schubert, D., Humphreys, S., Vitry, F., and Jacob, F., Induced differentiation of a neuroblastoma, *Dev. Biol.* **52**:514–546, 1971.

43. Prasad, K. N., Neuroblastoma clones: Prostaglandin versus dibutyryl cyclic AMP, 8-benzylthio-cyclic AMP, phosphodiesterase inhibitors and x-ray, *Proc. Soc. Exp. Biol. Med.* **140**:126–129, 1972.

44. Kimhi, Y., Palfrey, C., Spector, I., Barak, Y., and Littauer, U. Z., Maturation of neuroblastoma cells in the presence of dimethylsulfoxide. *Proc. Natl. Acad. Sci. U.S.A.* **73**:462–466, 1976.

45. Kirkland, W. L., and Burton, P. R., Cyclic adenosine monophosphate mediated stabilization of mouse neuroblastoma cell neurite microtubules exposed to low temperature, *Nature (London), New Biol.* **240**:205–207, 1972.

46. Helson, L., and Biedler, J. L., Catecholamines in neuroblastoma cells from human bone marrow, tissue culture and murine C-1300 tumor, *Cancer* **31**:1087–1091, 1973.

47. Herschman, H. R., and Lerner, M. P., Production of a nervous-system specific protein (14-3-2) by human neuroblastoma cells in culture, *Nature (London), New Biol.* **241**:242–244, 1973.

48. Prasad, K. N., and Gilmer, K. N., Demonstration of dopamine-sensitive adenylate cyclase in malignant neuroblastoma cells and change in sensitivity of adenylate cyclase to catecholamines in "differentiated" cells, *Proc. Natl. Acad. Sci. U.S.A.* **71**:2525–2529, 1974.

49. Prasad, K. N., Gilmer, K. N., Sahu, S. K., and Becker, G., Effect of neurotransmitters, guanosine triphosphate and divalent ions on the regulation of adenylate cyclase activity in malignant and adenosine cyclic 3':5'-monophosphate-induced differentiated neuroblastoma cells, *Cancer Res.* **35**:77–81, 1975.

50. Kimes, B., Tarikas, H., and Schubert, D., Neurotransmitter synthesis by two

clonal nerve cell lines: changes with culture growth and morphological differentiation, *Brain Res.* **79**:291–295, 1974.

51. Levi-Montalcini, R., and Angeletti, P. U., Essential role of nerve growth factor in survival and maintenance of dissociated sensory and sympathetic embryonic nerve cells *in vitro*, *Dev. Biol.* **7**:653–659, 1963.

52. Levi-Montalcini, R., and Angeletti, P. U., Nerve growth factor, *Physiol. Rev.* **48**:534–569, 1968.

53. Roisen, F., J., Murphy, R. A., Pichichero, M. E., and Braden, W. G., Cyclic adenosine monophosphate stimulation of axonal elongation, *Science* **175**:73–74, 1972.

54. Haas, D. C., Hier, D. B., Arnason, B. G. W., and Young, M., On a possible relationship of cyclic AMP to the mechanism of action of nerve growth factor, *Proc. Soc. Exp. Biol. Med.* **140**:45–47, 1972.

55. Nikodijevic, B., Nikodijevic, O., Yu, M. W., Pollard, H., and Guroff, G., The effect of nerve growth factor on cyclic AMP levels in superior cervical ganglia of the rat, *Proc. Natl. Acad. Sci. U.S.A.* **72**:4769–4771, 1975.

56. Ross, J., Olmsted, J. B., and Rosenbaum, J. L., The ultrastructure of mouse neuroblastoma cell in tissue culture. *Tissue Cell* **7**:107–136, 1975.

57. Chang, C. M., and Goldman, R. D., The localization of actin-like fibres in cultured neuroblastoma cells as revealed by heavy meromyosin binding, *J. Cell Biol.* **57**:867–874, 1973.

58. Heuser, J. E., and Reese, T. S., Evidence for recycling of synaptic vesicle membrane during transmitter release at the frog neuromuscular junction, *J. Cell Biol.* **57**:315–344, 1973.

59. Augusti-Tocco, G., Sato, G., Claude, P., and Potter, D., Clonal cell lines of neurons, in: *Control Mechanisms in the Expression of Cellular Phenotypes* (H. A. Padykula, ed.), pp. 109–120, Academic Press, New York, 1970.

60. Breakefield, X. O., Neale, E. A., Neale, J. H., and Jacobowitz, D. M., Localized catecholamine storage associated with granules in murine neuroblastoma cells, *Brain Res.* **92**:237–256, 1975.

61. Peters, A., Palay, S. L., and Webster, H. De F., *Fine Structure of the Nervous System*, p. 198, Harper and Row, New York, 1970.

62. Chalazonitis, A., and Greene, L. A., Enhancement in excitability properties of mouse neuroblastoma cell cultured in the presence of dibutyryl cyclic AMP, *Brain Res.* **72**:340–345, 1974.

63. Nelson, P., Christian, C., and Nirenberg, M., Synapse formation between clonal neuroblastoma x glioma hybrid cells and striated muscle cells, *Proc. Natl. Acad. Sci. U.S.A.* **73**:123–127, 1976

64. Redfern, P. A., Neuromuscular transmission in newborn rats, *J. Physiol. (London)* **209**:701–709, 1970.

65. Bennet, M. R., and Pettigrew, A. G., The formation of synapses in striated muscle during development, *J. Physiol. (London)* **241**:515–545, 1974.

66. Diamond, J., and Miledi, R., A study of foetal and new-born rat muscle fibres, *J. Physiol. (London)* **162**:393–408, 1962.

67. Fischbach, G. D. Synapse formation between dissociated nerve and muscle cells in low density cell cultures, *Dev. Biol.* **28**:407–429, 1972.

68. Steinbach, J. H., Harris, A. J., Patrick, J., Schubert, D., and Heinemann, S., Nerve–muscle interaction *in vitro*. Role of acetylcholine, *Gen. Physiol.* **62**:255–270, 1973.

69. Sytkowski, A. J., Vogel, Z., and Nirenberg, M. W., Development of acetylcholine receptor clusters on cultured muscle cells, *Proc. Natl. Acad. Sci. U.S.A.* **70**:270–274, 1973.

70. Fischbach, G. D., L. Cohen, S. A., The distribution of acetylcholine sensitivity over uninnervated and innervated muscle fibers grown in cell culture, *Dev. Biol.* **31**:147–162, 1973.

71. Landmesser, L., Contractile and electrical responses of vagus innervated frog sartorius muscles, *J. Physiol. (London)* **213**:707–725, 1971.

72. Nurse, C. A., and O'Lague, P. H., Formation of cholinergic synapses between dissociated sympathetic neurons and skeletal myotubes of the rat in cell culture, *Proc. Natl. Acad. Sci. U.S.A.* **72**:1955–1959, 1975.

73. Prasad, K. N., and Sheppard, J. R., Neuroblastoma cell culture: Membrane changes during cyclic AMP-induced morphological differentiation, *Proc. Soc. Exp. Biol. Med.* **141**:240–243, 1972.

74. Glick, M. C., Kimhi, Y., and Littauer, U. Z., Glycopeptides from surface membranes of neuroblastoma cells. *Proc. Natl. Acad. Sci. U.S.A.* **70**:1682–1687, 1973.

75. Truding, R., Shelanski, M. L., Daniels, M. P., and Morell, P., Comparison of surface membranes isolated from cultured murine neuroblastoma cells in the differentiated or undifferentiated state, *J. Biol. Chem.* **249**:3973–3982, 1974.

76. Brown, J. C., Surface glycoprotein characteristic of the differentiated state of neuroblastoma C-1300 cells, *Exp. Cell Res.* **69**:440–442, 1972.

78. Lazo, J. S., Prasad, K. N., and Ruddon, R. W., Synthesis and phosphorylation of chromatin-associated proteins in cAMP-induced "differentiated" neuroblastoma cells in culture, *Exp. Cell Res.* **100**:41–46, 1976.

79. Ehrlich, Y. H., Brunngraber, E. G., Sinha, P. K., and Prasad, K. N., Specific alterations in phosphorylation of cytosol proteins from differentiating neuroblastoma cells grown in culture, *Nature (London)* **265**:238–241, 1977.

80. Prasad, K. N., Mandal, B., Waymire, J. C., Lees, G. J., Vernadakis, A., and Weiner, N., Basal level of neurotransmitters synthesizing enzymes and effect of cyclic AMP agents on morphological differentiation of isolated neuroblastoma clones, *Nature (London), New Biol.* **241**:117–119, 1973.

81. Prasad, K. N., Gilmer, K., and Kumar, S., Morphologically "differentiated" mouse neuroblastoma cells induced by non-cyclic AMP agents: Level of cyclic AMP, nucleic acid and protein, *Proc. Soc. Exp. Biol. Med.* **143**:1168–1171, 1973.

82. Prasad, K. N., Waymire, J. C., and Weiner, N. A., Further study on the morphology and biochemistry of x-ray and dibutyryl cyclic AMP-induced "differentiated" neuroblastoma cells in culture, *Exp. Cell Res.* **74**:110–114, 1972.

83. Richelson, E., Stimulation of tyrosine hydroxylase activity in an adrenergic clone of mouse neuroblastoma by dibutyryl cyclic AMP, *Nature (London) New Biol.* **242**:175–177, 1973.

84. Orenberg, E. K., Vandenberg, S. R., Barchas, J. D., and Herman, M. M., Neurochemical studies in a mouse teratoma with neuroepithelial differentiation. Presence of cyclic AMP, serotonin and enzymes of the serotonergic, adrenergic and cholinergic systems, *Brain Res.* **101**:273–281, 1976.

85. Richelson, E., and Thompson, E. J., Transport of neurotransmitter precursors into cultured cells, *Nature (London), New Biol.* **241**:201–204, 1973.

86. Wexler, B., and Katzman, R., Effect of dibutyryl cyclic AMP and dexamethasone on noradrenaline synthesis in isolated superior cervical ganglia, *J. Neurochem.* **22**:5–10, 1974.

87. Culver, B., Sahu, S. K., Vernadakis, A., and Prasad, K. N., Effects of 5-

(3,3-dimethyl-1-triazeno) imidazole-4-carboxamide (NSC 45388, DTIC) on neuroblastoma cells in culture, *Biochem. Biophys. Res. Commun.* **76**:778–783, 1977.

88. Mackay, A. V. P., and Iversen, L. I., Increased tyrosine hydroxylase activity of sympathetic ganglia cultured in the presence of dibutyryl cyclic AMP, *Brain Res.* **48**:424–426, 1972.

89. Anagnoste, B., Shirron, C., Friedman, E., and Goldstein, M., Effect of dibutyryl cyclic adenosine monophosphate on $^{14}$C-dopamine biosynthesis in rat brain striatal slices. *J. Pharmacol. Exp. Ther.* **191**:370–376, 1974.

90. Goldstein, M., Bronaugh, R. L., Ebstein, B., and Roberge, C., Stimulation of tyrosine hydroxylase activity by cyclic AMP in synaptosomes and in soluble striatal enzyme preparations, *Brain Res.* **109**:563–574, 1976.

91. Anagnoste, B., Freedman, L. S., Goldstein, M., Broome, J., and Fuxe, K., Dopamine β-hydroxylase activity in mouse neuroblastoma tumors and in cell cultures, *Proc. Natl. Acad. Sci. U.S.A.* **69**:1883–1886, 1972.

92. Hamprecht, B., Traber, J., and Lamprecht, F., Dopamine β-hydroxylase activity in cholinergic neuroblastoma x glioma hybrid cells; increase of activity by $N^6O_2'$-dibutyryl adenosine 3':5'-cyclic monophosphate, *FEBS Lett.* **42**:221–226, 1974.

93. Keen, P., and McLean, W. G., Effect of dibutyryl cyclic AMP on levels of dopamine β-hydroxylase in isolated superior cervical ganglia, *Arch. Pharmacol.* **275**:465–469, 1972.

94. Keen, P., and McLean, W. G., Effect of dibutyryl cyclic AMP and dexamethasone on noradrenaline synthesis in isolated superior cervical ganglia, *J. Neurochem.* **22**:5–10, 1974.

95. Molinoff, P. B., and Axelrod, J., Biochemistry of catecholamines, *Ann. Rev. Biochem.* **40**:465–500, 1971.

96. Thoenen, H., Neuronally mediated enzyme induction in adrenergic neurons and adrenal chromaffin cells, *Biochem. Soc. Symp.* **36**:3–15, 1972.

97. Thoenen, H., Mueller, R. A., and Axelrod, J., Increased tyrosine hydroxylase activity after drug-induced alteration of sympathetic transmission, *Nature (London)* **221**:1264, 1969.

98. Molinoff, P. B., Brimijoin, S., Weinshilboum, R., and Axelrod, J., Neurally mediated increase in dopamine-β-hydroxylase activity, *Proc. Natl. Acad. Sci. U.S.A.* **66**:453–458, 1970.

99. Mueller, R. A., Thoenen, H., and Axelrod, J., Inhibition of neuronally induced tyrosine hydroxylase by nicotinic receptor blockade, *Eur. J. Pharmacol.* **10**:51–56, 1970.

100. Patrick, R. L., and Kirshner, N., Acetylcholine-induced stimulation of catecholamine recovery in denervated rat adrenal after reserpine-induced depletion, *Mol. Pharmacol.* **7**:389–396, 1971.

101. Guidotti, A., and Costa, E., Involvement of adenosine 3',5'-monophosphate in the activation of tyrosine hydroxylase elicited drugs, *Science* **179**:902–904, 1973.

102. McAfee, D. A., Schorderet, M., and Greengard, P., Adenosine 3',5'-monophosphate in nervous tissue: Increase associated with synaptic transmission, *Science* **171**:1156–1158, 1971.

103. Greengard, P., and McAfee, D. A., Adenosine 3'-cyclic monophosphate as a mediator in the action of neurohumoral agents, *Biochem. Soc. Symp.* **36**:87–102, 1972.

104. Costa, E., and Guidotti, A., The role of 3',5'-cyclic adenosine monophosphate in the regulation of adrenal medullary function, in: *New Concepts in Neurotransmitter Regulation* (A. J. Mandell, ed.), pp. 135–152, Plenum Press, New York, 1973.

105. Goodman, R., Oesch, F., and Thoenen, H., Changes in enzyme patterns pro-
     duced by high potassium concentration and diburyryl cyclic AMP in organ cul-
     tures of sympathetic ganglia, *J. Neurochem.* **23**:369–378, 1974.
106. Otten, U., Mueller, R. A., Oesch, F., and Thoenen, H., Location of an
     isoproterenol-responsive cyclic AMP pool in adrenergic nerve cell bodies and its
     relationship to tyrosine 3-monooxygenase induction, *Proc. Natl. Acad. Sci. U.S.A.*
     **71**:2217–2221, 1974.
107. Prasad, K. N., and Mandal, B., Choline acetyltransferase level in cyclic AMP and
     x-ray induced morphologically differentiated neuroblastoma cells in culture,
     *Cytobios* **8**:75–80, 1973.
108. Simantov, R., and Sachs, L., Enzyme regulation in neuroblastoma cells, selection
     of clones with low acetylcholinesterase activity and the independent control of
     acetylcholinesterase and choline-O-acetyl-transferase, *Eur. J. Biochem.* **30**:123–
     129, 1972.
109. Rosenberg, R. N., Vandeventer, L., De Francesco, L., and Friedkin, M. E., Regu-
     lation of the synthesis of choline-O-acetyltransferase and thymidylate synthetase
     in mouse neuroblastoma in cell culture, *Proc. Natl. Acad. Sci. U.S.A.* **68**:1436–1440,
     1971.
110. Prasad, K. N., and Mandal, B., Catechol-o-methyl-transferase activity in di-
     butyryl cyclic AMP, prostaglandin and x-ray-induced differentiated neuroblas-
     toma cell culture, *Exp. Cell Res.* **74**:532–534, 1972.
111. Prasad, K. N., and Vernadakis, A., Morphological and biochemical study in x-ray
     and dibutyryl cyclic AMP-induced differentiated neuroblastoma cells, *Exp. Cell
     Res.* **70**:27–32, 1972.
112. Lanks, K. W., Turnbull, J. D., Aloyo, V. J., Dorwin, J. M., and Papirmeister, B.,
     Sulfur mustards induce neurite extension and acetylcholinesterase synthesis in
     cultured neuroblastoma cell, *Exp. Cell Res.* **93**:355–362, 1975.
113. Ruffner, B. W., and Smith, M., Biochemical differentiation of a murine
     ganglioneuroblastoma in tissue culture, *Exp. Cell Res.* **89**:442–447, 1974.
114. Schneider, F. H., Effect of sodium butyrate on mouse neuroblastoma cells in
     culture, *Biochem. Pharmacol.* **25**:2309–2317, 1976.
115. Lanks, K. W., Dorwin, J. M., and Papirmeister, B., Increased rate of acetyl-
     cholinesterase synthesis in differentiating neuroblastoma cells, *J. Cell Biol.*
     **63**:824–830, 1974.
116. Cox, G. C., and Juniper, B. E., Autoradiographic evidence for paramural-body
     function, *Nature (London) New Biol.* **243**:116–117, 1973.
117. Simantov, R., and Sachs, L., Regulation of acetylcholine receptors in relation to
     acetylcholinesterase in neuroblastoma cells, *Proc. Natl. Acad. Sci. U.S.A.* **70**:2902–
     2905, 1973.
118. Simantov, R., and Sachs, L., Different mechanisms for the induction of acetyl-
     cholinesterase in neuroblastoma cells, *Dev. Biol.* **45**:382–385, 1975.
119. Ruffner, B. W., and Grieshaber, D. M., Biochemical differentiation of a murine
     neuroblastoma *in vitro* and *in vivo*, *Cancer Res.* **34**:551–558, 1974.
120. Harkins, J., Arsenault, M., Schlesinger, K., and Kates, J., Induction of neuronal
     functions: Acetylcholine-induced acetylcholinesterase activity in mouse neuro-
     blastoma cells, *Proc. Natl. Acad. Sci. U.S.A.* **69**:3161–3164, 1972.
121. LaBrosse, E. H., and Karon, M., Catechol-O-methyltransferase activity in neuro-
     blastoma tumour, *Nature (London)* **196**:1222–1223, 1962.
122. Blume, A., Gilbert, F., Wilson, S., Farber, J., Rosenberg, R., and Nirenberg, M.,

Regulation of acetylcholinesterase in neuroblastoma cells, *Proc. Natl. Acad. Sci. U.S.A.* **67**:786–792, 1970.

123. Basu, S., Moskal, J. R., and Gardner, D. A., Scanning electronmicroscopic and glycosphingolipid biosynthetic studies of differentiating mouse neuroblastoma cells, in: *Ganglioside Function: Biochemical and Pharmacological Implications* (G. Porcellati, B. Ceccarelli, and G. Tettamanti, eds.), pp. 45–63, Plenum Press, New York, 1976.

124. Moskal, J. R., Gardner, D. A., and Basu, S., Changes in glycolipid glycosyltransferases and glutamate decarboxylase and their relationship to differentiation in neuroblastoma cells, *Biochem. Biophys. Res. Commun.* **61**:751–758, 1974.

125. Phillipson, O. T., and Sandler, M., The influence of nerve growth factor, potassium depolarization and dibutyryl (cyclic) adenosine 3′,5′-monophosphate on explant cultures of chick embryo sympathetic ganglia, *Brain Res.* **90**:273–281, 1975.

126. Schimizu, H., Creveling, C. R., and Daly, J. W., Effect of membrane depolarization and biogenic amines on the formation of cyclic AMP in incubated brain slices, in: *Role of Cyclic AMP in Cell Function*, Vol. 3 (P. Greengard and E. Costa, eds.), pp. 135–154, Raven Press, New York, 1970.

127. Cohen, S. S., *Introduction to Polyamines*, 179 pp., Prentice Hall, Englewood Cliffs, N.J., 1971.

128. Bachrach, U., Cyclic AMP-mediated induction of ornithine decarboxylase of glioma and neuroblastoma cells, *Proc. Natl. Acad. Sci. U.S.A.* **72**:3087–3091, 1975.

129. Anderson, T. R., and Schanberg, S. M., Ornithine decarboxylase activity in developing rat brain, *J. Neurochem.* **19**:1471–1478, 1972.

130. Sturman, J. A., and Gaull, G. E., Polyamine biosynthesis in human fetal liver and brain, *Pediat. Res.* **8**:231–237, 1974.

131. Bachrach, U., Induction of ornithine decarboxylase in glioma and neuroblastoma cells, *FEBS Lett.* **68**:63–67, 1976.

132. Nissen, C., Ciesielski-Treska, J., Hertz, L., and Mandel, P., Regulation of oxygen consumption in neuroblastoma cells. Effects of differentiation and of potassium, *J. Neurochem.* **20**:1029–1035, 1973.

133. Nissen, C., Ciesielski-Treska, J., Hertz, L., and Mandel, P., Rates of oxygen uptake in proliferating and differentiating neuroblastoma cells, *Brain Res.* **39**:264–267, 1972.

134. Prasad, K. N., Sahu, S. K., and Kumar, S., Relationship between cyclic AMP level and differentiation of neuroblastoma cells in culture, in: *Differentiation and Control of Malignancy of Tumor Cells* (W. Nakahara, T. Ono, T. Sugimura, and H. Sugano, eds.), pp. 287–309, University of Tokyo Press, Tokyo, 1974.

135. Ciesielski-Treska, J., Tholey, G., Wurtz, B., and Mandel, P., Enzymic modifications in a cultivated neuroblastoma clone after bromodeoxyuridine treatment, *J. Neurochem.* **26**:465–469, 1976.

136. Criss, W. E., A review of isozymes in cancer, *Cancer Res.* **31**:1523–1542, 1971.

137. Dawson, G., Kemp, S. F., Stoolmiller, A. C., and Dorfman, A., Biosynthesis of glycosphingolipids by mouse neuroblastoma (NB41A) rat glia (RGC-6) and human glia (CHB-4) in cell culture, *Biochem. Biophys. Res. Commun.* **44**:687–694, 1971.

138. Yogeeswaran, G., Murray, R. K., Pearson, M. L., Sanwal, B. D., McMorris, F. A., and Ruddle, F. H., Glysosphingolipids of clonal lines of mouse neuroblastoma and neuroblastoma × L-cell hybrids *J. Biol. Chem.* **248**:1231–1239, 1973.

139. Basu, M., and Basu, S., Enzymatic synthesis of a tetraglycosylceramide by a galactosyltransferase from rabbit bone marrow, *J. Biol. Chem.* **247**:1489–1495, 1972.

140. Sarlième, L. L., Neskovic, N. M., Freysz, L., Mandel, P., and Rebel, G., Ceramide galactosyltransferase and cerebroside sulphotransferase in chicken brain cellular fractions and glial and neuronal cells in culture, *Life Sci.* **18**:251–269, 1976.

141. Schnebli, H. P., and Burger, M. M., Selective inhibition of growth of transformed cells by protease inhibitors, *Proc. Natl. Acad. Sci. U.S.A.* **69**:3825–2827, 1972.

142. Wachsman, J. T., and Biedler, J. L., Fibrinolytic activity associated with human neuroblastoma cells, *Exp. Cell Res.* **86**:264–268, 1974.

143. Laug, W. E., Jones, P. A., Nye, C. A., and Benedict, W. F., The effect of cyclic AMP and prostaglandins on the fibrinolytic activity of mouse neuroblastoma cells, *Biochem. Biophys. Res. Commun.* **68**:114–119, 1976.

144. Prasad, K. N., Kumar, S., Gilmer, K., and Vernadakis, A., Cyclic AMP-induced differentiated neuroblastoma cells: Changes in total nucleic acid and protein contents, *Biochem. Biophys. Res. Commun.* **50**:973–977, 1973.

145. Zucco, F., Persico, M., Felsani, A., Metafora, S., and Augusti-Tocco, G., Regulation of protein synthesis at the translational level in neuroblastoma cells. *Proc. Natl. Acad. Sci. U.S.A.* **72**:2289–2293, 1975.

146. Augusti-Tocco, G., Casola, L., and Romano, M., RNA metabolism in neuroblastoma cultures. II. Synthesis of nonribosomal RNA, *Cell Differ.* **3**:313–320, 1974.

147. Casola, L., Romano, M., and Dimatteo, G., Augusti-Tocco, G., and Estenoz, M., RNA metabolism in neuroblastoma cultures. I. Ribosomal RNA, *Dev. Biol.* **41**:371–379, 1974.

148. Judes, C., Sensenbrenner, M., Jacob, M., and Mandel, P., Differentiation of chick embryo cerebral hemispheres. I. Incorporation of tritiated uridine into the acid-soluble nucleotide pool and into total RNA *in vitro*, *Brain Res.* **51**:241–251, 1974.

149. Judes, C., and Jacob, M., Differentiation of chick embryo cerebral hemispheres. II. Incorporation of ³H-uridine into 2S, S, 18S, 5S and 4S RNA, *Brain Res.* **51**:253–267, 1973.

150. Lim, R., and Mitsunobu, K., Effect of dibutyryl cyclic AMP in nucleic acid and protein synthesis in neuronal and glial tumor cells, *Life Sci.* **11**:1063–1070, 1972.

151. Glazer, R. I., and Schneider, F. H., Effect of adenosine 3′:5′-monophosphate and related agents on ribonucleic acid synthesis and morphological differentiation in mouse neuroblastoma cells in culture, *J. Biol. Chem.* **250**:2745–2749, 1975.

152. Bondy, S. C., Prasad, K. N., and Purdy, J. L., Neuroblastoma: Drug-induced differentiation increases proportion of cytoplasmic RNA that contains polyadenylic acid, *Science* **186**:359–361, 1974.

153. Prasad, K. N., Bondy, S. C., and Purdy, J. L., Polyadenylic acid containing cytoplasmic RNA increases in adenosine 3′,5′-cyclic monophosphate-induced "differentiated" neuroblastoma cells in culture, *Exp. Cell Res.* **94**:388–394, 1975.

154. Sarkar, P. K., Goldman, B., and Moscona, A. A., Involvement of poly-A in selective gene expression: Suppression of enzyme induction in neural retina by inhibitors of poly-A synthesis, *Biochem. Biophys. Res. Commun.* **50**:308–315, 1973.

155. Prasad, K. N., and Sinha, P. K., Effect of sodium butyrate on mammalian cells in culture. A review, *In Vitro*, **12**:125–132, 1975.

156. Weber, G., The molecular correlation concept of neoplasia and the cyclic AMP system, in: *The Role of Cyclic Nucleotides in Carcinogenesis* (J. Schultz and H. G. Gratzner, eds.), pp. 57–94, Academic Press, New York, 1973.

157. Kernell, A. M., Bolund, L., and Ringertz, N. R., Chromatin changes during erythropoiesis, *Exp. Cell Res.* **65**:1–6, 1971.

158. Littauer, U. Z., Schmitt, H., and Gozes, T., Properties and synthesis of tubulin in neuroblastoma cells, *J. Natl. Cancer Inst.* **57**:647-651, 1976.

159. Balhorn, R., Bordwell, J., Sellers, L., Granner, D., and Chalkley, R., Histone phosphorylation and DNA synthesis are linked in synchronous cultures of HTC cells, *Biochem. Biophys. Res. Commun.* **46**:1326-1333, 1972.

160. Gurley, L. R., Walters, R. A., and Tobey, R. A., The metabolism of histone fractions. IV. Synthesis of histones during the $G_1$-phase of the mammalian life cycle, *Arch. Biochem. Biophys.* **148**:633-641, 1972.

161. Gurley, L. R., Walters, R. A., and Tobey, R. A., Cell cyclic-specific changes in histone phosphorylation associated with cell proliferation and chromosome condensation, *J. Cell Biol.* **60**:356-364, 1974.

162. Krause, M. O., and Inasi, B. S., Histones from exponential and stationary L-cells. Evidence for metabolic heterogeneity of histone fractions retained after isolation of nuclei, *Arch. Biochem. Biophys.* **164**:179-184, 1974.

163. Burdman, J. A., The relationship between DNA synthesis and the synthesis of nuclear proteins in rat brain during development, *J. Neurochem.* **19**:1459-1469, 1972.

164. Fujitani, H., and Holoubek, V., Nonhistone nuclear proteins of rat brain, *J. Neurochem.* **23**:1215-1224, 1974.

165. Olpe, H. R., Van Hahn, H. P., and Honegger, C. G., The non-histone protein pattern of rat brain during ontogenesis, *Experientia* **29**:665-666, 1972.

166. Elgin, S. C. R., Boyd, J. B., Hood, L. E., Wray, W., and Wu, F. C., A prologue to the study of the nonhistone chromosomal proteins, *Cold Spring Harbor Symp. Quant. Biol.* **38**:821-833, 1973.

167. Zornetzer, M. S., and Stein, G. S., Gene expression in mouse neuroblastoma cells: Properties of the genome, *Proc. Natl. Acad. Sci. U.S.A.* **72**:3119-3123, 1975.

168. Olmsted, J. B., Carlson, K., Klebe, R., Ruddle, F., and Rosenbaum, J., Isolation of microtubule protein from cultured mouse neuroblastoma cells, *Proc. Natl. Acad. Sci. U.S.A.* **65**:129-136, 1970.

169. Solomon, F., Monard, D., and Rentsch, M., Stabilization of colchicine-binding activity of neuroblastoma, *J. Mol. Biol.* **78**:569-573, 1973.

170. Wiche, G., Zomzely-Neurath, C., and Blume, A. J., *In vitro* synthesis of mouse neuroblastoma tubulin, *Proc. Natl. Acad. Sci. U.S.A.* **71**:1460-1450, 1974.

171. Littauer, U. Z., Schmitt, H., and Gozes, T., Properties and synthesis of tubulin in neuroblastoma cells, *Natl. Cancer Inst.* **57**:647-651, 1976.

172. Miller, C., and Kuehl, W. M., Isolation and characterization of myosin from cloned rat glioma cells and mouse neuroblastoma cells, *Brain Res.* **108**:115-124, 1976.

173. Burton, P. R., and Kirkland, W. L., Actin detected in mouse neuroblastoma cells by binding of heavy meromyosin, *Nature (London) New Biol.* **239**:244-246, 1972.

174. Lessard, J. L., Goldblatt, D., Rein, D., and Carlton, D., Localization of actin in neuroblastoma cells by immunofluorescence, *J. Cell Biol.* **70**:150a, 1976.

175. Augusti-Tocco, G., Casola, L., and Grasso, A., Neuroblastoma cells and 14-3-2. A brain specific protein, *Cell Differ.* **2**:157-161, 1973.

176. Penit, J., Huot, J., and Jard, S., Neuroblastoma cell adenylate cyclase: Direct activation by adenosine and prostaglandins, *J. Neurochem.* **26**:265-273, 1976.

177. Blume, A. J., and Foster- C. J., Mouse neuroblastoma cell adenylate cyclase: Regulation by 2-chloroadenosine, prostaglandin $E_1$ and the cations $Mg^{2+}$, $Ca^{2+}$ and $Mn^{2+}$, *J. Neurochem.* **26**:305-311, 1976.

178. Blume, A. J., and Foster, C. J., Mouse neuroblastoma adenylate cyclase.

Adenosine and adenosine analogues as potent effectors of adenylate cyclase activity, *J. Biol. Chem.* **250**:5003–5008, 1975.

179. Prasad, K. N., Gilmer, K. N., and Sahu, S. K., Demonstration of acetylcholine-sensitive adenyl cyclase in malignant neuroblastoma cells in culture, *Nature (London)* **249**:765–767, 1974.

180. Blume, A. J., and Foster, C. J., Neuroblastoma adenylate cyclase. Role of 2-chloroadenosine, prostaglandin E$_1$, and guanine nucleotides in regulation of activity, *J. Biol. Chem.* **251**:3399–3404, 1976.

181. Levey, G. S., Solubilization of myocardial adenyl cyclase, *Biochem. Biophys. Res. Commun.* **38**:86–92, 1970.

182. DeHaen, C., A new kinetic analysis of the effects of hormones and fluoride ion, *J. Biol. Chem.* **249**:2756–2762, 1974.

183. Harris, A. J., and Dennis, M. J., Acetylcholine sensitivity and distribution on mouse neuroblastoma cells, *Science* **167**:1253–1255, 1970.

184. Gilman, A. G., and Nirenberg, M. W.- Regulation of adenosine 3′,5′-cyclic monophosphate metabolism in cultured neuroblastoma cells, *Nature (London)* **234**:356–357, 1971.

185. Blume, A. J., Dalton, C., and Sheppard, H., Adenosine-mediated elevation of cyclic 3′:5′-adenosine monophosphate concentrations in cultured mouse neuroblastoma cells, *Proc. Natl. Acad. Sci. U.S.A.* **70**:3099–3102, 1973.

186. Prasad, K. N., Kumar, S., Becker, G., and Sahu, S. K., The role of cyclic nucleotides in differentiation of neuroblastoma cells in culture, in: *Cyclic Nucleotides in Diseases* (B. Weiss, ed), pp. 45–66, University Park Press, Baltimore, 1975.

187. Sahu, S. K., and Prasad, K. N., Effect of neurotransmitters and prostaglandin E$_1$ on cyclic AMP levels in various clones of neuroblastoma cells in culture, *J. Neurochem.* **24**:1267–1269, 1975.

188. Schubert, D., Tarikas, H., and Lacorbiere, M., Neurotransmitter regulation of adenosine 3′,5′-monophosphate on clonal nerve, glia and muscle cell lines, *Science* **192**:471–472, 1976.

189. Blume, A. J., Foster, C. J., and Karp, G., Acetylcholine inhibition of adenosine and prostaglandin E$_1$ stimulants of mouse neuroblastoma cAMP levels. Presented at the fifth meeting of the American Society for Neurochemistry. New Orleans, p. 150, 1974.

190. Traber, J., Fischer, K., Latzin, S., and Hamprecht, B., Morphine antagonises action of prostaglandin in neuroblastoma and neuroblastoma × glioma hybrid cells, *Nature (London)* **253**:120–122, 1975.

191. Prasad, K. N., and Kumar, S., Role of cyclic AMP in differentiation of human neuroblastoma cells in culture, *Cancer* **36**:1338–1343, 1975.

192. Prasad, K. N., Sahu, S. K., and Sinha, P. K., Cyclic nucleotides in the regulation of expression of differentiated functions in neuroblastoma cells, *J. Natl. Cancer Inst.* **57**:619–631, 1976.

193. Matsuzawa, H., and Nirenberg, M., Receptor-mediated shifts in cGMP and cAMP levels in neuroblastoma cells, *Proc. Natl. Acad. Sci. U.S.A.* **72**:3472–3476, 1975.

194. Prasad, K. N., Becker, G., and Tripathy, K., Differences and similarities between guanosine 3′,5′-cyclic monophosphate phosphodiesterase and adenosine 3′,5′-cyclic monophosphate phosphodiesterase activities in neuroblastoma cells in culture, *Proc. Soc. Exp. Biol. Med.* **149**:757–762, 1975.

195. Sinha, P. K., and Prasad, K. N., A further study on the regulation of cyclic

nucleotide phosphodiesterase activity in neuroblastoma cells. Effect of growth, *in Vitro* (in press).

196. Cheung, W. Y., Cyclic nucleotide phosphodiesterase, In: *Role of Cyclic AMP in Cell Function* (P. Greengard and E. Costa, eds.), pp. 51–65, Raven Press, New York, 1970.

197. Prasad, K. N., and Kumar, S., Cyclic 3',5'-AMP phosphodiesterase activity during cyclic AMP-induced differentiation of neuroblastoma cells in culture, *Proc. Soc. Exp. Biol. Med.* **142:**406–409, 1973.

198. Kumar, S., Becker, G., and Prasad, K. N., Cyclic adenosine 3':5'-monophosphate phosphodiesterase activity in malignant and cyclic adenosine 3',5'-monophosphate-induced "differentiated" neuroblastoma cells, *Cancer Res.* **35:**82–87, 1975.

199. Gill, G. N., and Garren, L. D., A cyclic-3',5'-adenosine monophosphate dependent protein kinase from the adrenal cortex: comparison with a cyclic AMP binding protein, *Biochem. Biophys. Res. Commun.* **39:**335–343, 1970.

200. Tao, M., Salas, M. L., and Lipmann, F., Mechanism of activation by adenosine 3':5'-cyclic monophosphate of a protein phosphokinase from rabbit reticulocytes, *Proc. Natl. Acad. Sci. U.S.A.* **67:**408–414, 1970.

201. Reimann, E. M., Brostrom, C. O., Corbin, J. D., King, C. A., and Krebs, E. G., Separation of regulatory and catalytic subunits of the cyclic 3',5'-adenosine monophosphate-dependent protein kinase(s) of rabbit skeletal muscle, *Biochem. Biophys. Res. Commun.* **42:**187–194, 1971.

202. Kumon, A., Yamamura, H., and Nishizuka, Y., Mode of action of adenosine 3',5'-cyclic phosphate on protein kinase from rat liver, *Biochem. Biophys. Res. Commun.* **41:**1290–1297, 1970.

203. Prasad, K. N., Sinha, P. K., Sahu, S. K., and Brown, J. L., Binding of cyclic nucleotoides with soluble proteins increases in differentiated neuroblastoma cells in culture, *Biochem. Biophys. Res. Commun.* **66:**131–138, 1975.

204. Prasad, K. N., Sinha, P. K., Sahu, S. K., and Brown, J. L., Binding of cyclic nucleotides with proteins in malignant and adenosine 3':5'-cyclic monophosphate-induced "differentiated" neuroblastoma cells in culture, *Cancer Res.* **36:**2290–2296, 1976.

205. Prashad, N., and Rosenberg, R. N., Phosphorylation of proteins by dibutyryl cAMP during differentiation of mouse neuroblastoma cells, Presented at the 8th annual meeting of the American Society for Neurochemistry, Denver, Colorado, March 13–18, 1977.

206. Butcher, R. W., and Sutherland, E. W., Adenosine 3',5'-phosphate in biological materials. Purification and properties of cyclic 3',5'-nucleotide phosphodiesterase and use of this enzyme to characterize adenosine 3',5'-phosphate in human urine, *J. Biol. Chem.* **237:**1244–1255, 1962.

207. Ebadi, M. S., Weiss, B., and Costa, E., Distribution of cyclic adenosine monophosphate in rat brain, *Arch. Neurol.* **24:**353–357, 1971.

208. Sutherland, E. W., Rall, T. W., and Menon, T., Adenylate cyclase. I. distribution, preparation, and properties, *J. Biol. Chem.* **237:**1220–1227, 1962.

209. Prasad, K. N., Fogleman, D., Gaschler, M., Sinha, P. K., and Brown, J. L., Cyclic nucleotide-dependent protein kinase activity in malignant and cyclic AMP-induced "differentiated" neuroblastoma cells in culture, *Biochem. Biophys. Res. Commun.* **68:**1248–1255, 1976.

210. Ehrlich, Y. H., and Routtenberg, A., Cyclic AMP regulates phosphorylation of

three protein components of rat cerebral cortex membranes for thirty minutes, *FEBS Lett.* **45:**237–243, 1974.

211. Seite, R., Leonetti, J., Luciani-Vuillet, J., and Vio, M., Cyclic AMP and ultrastructural organization of the nerve cell nucleus; stimulation of nuclear microtubules and microfilaments assembly in sympathetic neurons, *Brain Res.* **124:**41–51, 1977.

212. Prasad, K. N., Abnormal regulation of cyclic AMP phosphodiesterase: A hypothesis for the development of cancer of nerve cells, *Differentiation* **2:**367–369, 1974.

213. Knapp, S., and Mandell, A. J., Serotonin biosynthetic capacity of mouse C-1300 neuroblastoma cells in culture, *Brain Res.* **66:**547–551, 1974.

214. Prasad, K. N., Sinha, P. K., Ramanuzam, M., and Sakamoto, A., Sodium ascorbate potentiates the growth inhibitory effect of certain agents on neuroblastoma cells in culture, *Proc. Natl. Acad. Sci. U.S.A.* **76:**829–832, 1979.

215. Prasad, K. N., and Sinha, P. K., Regulation of differentiated functions and malignancy in neuroblastoma cells in culture, in: *Differentiation and Neoplasia* (G. F. Saunders, ed.), pp. 111–141, M. D. Anderson Hospital, Houston, 1978.

216. Orr, C. W. M., Studies on ascorbic acid. Factors influencing the ascorbate-mediated inhibition of catalase, *Biochemistry* **6:**2995–3001.

217. Helstroem, I. E., Helstroem, K. D., Pierce, G. E., and Bill, A. H., Demonstration of cell-bound and humoral immunity against neuroblastoma cells, *Proc. Natl. Acad. Sci. U.S.A.* **60:**1231–1238, 1968.

218. Imashuku, S., Todo, S., Amano, T., Mizukawa, K., Sugimoto, T., and Kusunoki, T., Cyclic AMP in neuroblastoma, glanglioneuroma and sympathetic ganglia, *Experientia* **33:**1507, 1977.

# ROLE OF AGENTS OTHER THAN cAMP IN THE REGULATION OF DIFFERENTIATION OF NERVE CELLS

## NERVE GROWTH FACTOR

While studying the effect of peripheral tissue on the developing nervous system, Bueker[1] observed a marked hypertrophy of chick embryonic sensory and sympathetic ganglia after the transplantation of mouse sarcoma tissue. It was later found[2] that this hypertrophy was due to the release of a humoral factor by the tumor explants, and this factor was named nerve growth factor (NGF). NGF is essential for the growth, development, and maintenance of embyronic sympathetic ganglia, dorsal root ganglia, and sensory ganglia *in vivo* and *in vitro*.[3-5] NGF appears to have a rather selective effect on the sympathetic system, although several reports indicate that NGF also stimulates other cell types; however, the magnitude of such effects is minor.[6] Therefore, the main physiological effect of NGF may be on the sympathetic nervous system.[6] This is supported by the fact that NGF antiserum produces a rapid and complete destruction of all para- and prevertebral ganglia.[7,8] The NGF-antiserum-induced morphological changes that eventually lead to cell death are different from those produced by cytotoxic drugs.[6] For example, the initial morphological changes seen after treatment with a cytotoxic agent such as 6-hydroxydopamine (6-OHDA) are first observed in the cytoplasm. These changes gradually spread to the nucleus and are followed by cell death. On the other hand, the lesion produced by NGF antiserum appears first in the nucleus and then in the cytoplasm. The nuclear damage is evidenced by nucleolar segregation and condensation of the chromatin.

The hypertrophy of sympathetic ganglia in response to NGF is due to an increase in both cell number and size. Electron-microscopic studies reveal that treatment with NGF produces significant changes in cellular ultrastructure,[9] with an enrichment of membranous constituents such as the endoplasmic reticulum and Golgi apparatus. The most characteristic feature of these cells is the appearance of large bundles of microtubules and microfilaments.[6]

NGF increases glucose oxidation, primarily via the direct oxidative pathway,[10] and increases the incorporation of acetate into lipid,[11] amino acid into protein,[12] and uridine into RNA.[12] The increase in RNA synthesis is particularly marked; therefore, it has been postulated[12] that NGF may act via stimulation of the synthesis of a new messenger RNA. However, this postulation was made doubtful by the observation that neurite outgrowth from NGF-stimulated sympathetic and dorsal root ganglia, cultured in a semisolid agar medium or in plasma clot, was not abolished following complete inhibition of RNA synthesis by actinomycin D.[13,14] A recent study,[15] using fetal calf serum-coated Falcon tissue culture plates in a liquid medium, showed that neurite outgrowth of chick embryo dorsal root and sympathetic ganglia induced by NGF depended upon RNA synthesis; however, neurite outgrowth no longer depended upon RNA synthesis if the ganglia were cultured in plasma clot, or on either collagen- or poly-L-lysine-coated plates. Thus, the type of growth substratum is an important factor in the mechanism involved in the expression of NGF-induced neurites. In contrast to RNA synthesis, NGF-induced neurite formation depends upon new protein synthesis, and this requirement for neurite outgrowth does not vary as a function of the type of substratum used.[12,15] Another study,[16] using a different protein hormone, has shown that stimulation of RNA synthesis is not a prerequisite for the expression of neurites. For example, insulin, structurally related to NGF, stimulates RNA synthesis in sympathetic ganglia without stimulating neurite outgrowth.[16]

The effects of NGF on biochemical processes unique to sympathetic neurons have also been investigated. NGF increases the activity of TH,[17,18] DBH,[17] and CAT.[19] An increase in catecholamine synthesis is also indicated by a large increase of histofluorescence in cells treated in vivo with NGF, suggesting that NGF causes functional stimulation of sympathetic neurons. The specificity of NGF on catecholamine-synthesizing enzymes of the sympathetic nervous system was demonstrated by comparing the activities of enzymes in sympathetic ganglia and adrenal medullary cells, which contain similar catecholamine synthesizing enzymes. NGF had virtually no effect on the activities of either TH or DBH in the adrenal medulla (Figure 4-1).

NGF increases MAO activity in chick embryo sympathetic ganglia[20] in a dose-dependent fashion, with maximum effects seen at a concentration of about 8 U/ml NGF. Peak activity compares closely with values seen in chick ganglia *in vivo;* however, in rat superior cervical ganglia *in vivo,* NGF does not increase MAO activity.[17] Because the ganglia in culture remain devoid of their presynaptic input, and because this input appears to have marked effects on the development of ganglia *in vivo,*[21,22] the influence of depolarizing concentrations (45 mM) of potassium on MAO activity was investigated. Potassium-induced depolarization also increases MAO activity[20]; however, the extent of increase is dependent upon NGF. The mechansims of action of NGF and potassium may be different. High concentrations of potassium increase primarily the cell number, but NGF increases both cell number and total protein.[20] NGF partially,

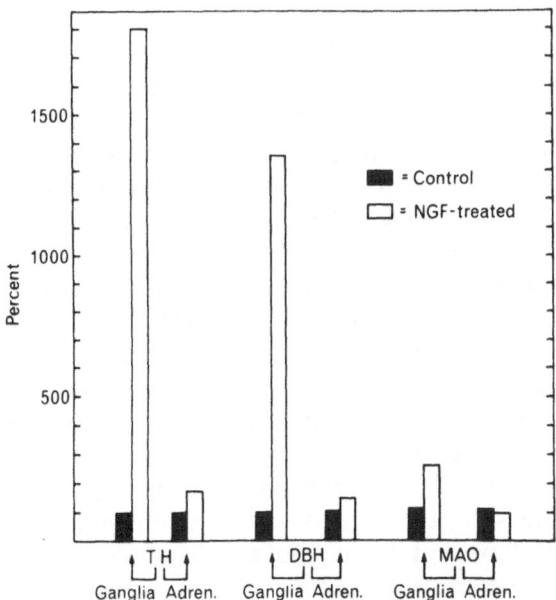

**Figure 4-1.** A comparison of the effects of nerve growth factor on the levels of TH and DBH in sympathetic ganglia and adrenal medullary tissue. (From Thoenen *et al.*[17])

but not fully, mimics the effect of depolarization on MAO activity.[20] A large dose of NGF can also partially replace the effect of deafferentation on TH activity.[22] These findings support the concept that depolarizing stimuli and NGF effect different processes in the ganglia. NGF may partly influence the neurons through its terminal and therefore may carry information from the center to the periphery.[23,24] Dibutyryl-cAMP mimics the effect of a potassium-induced increase in MAO activity in chick embryo sympathetic ganglia.[20] This is not surprising, because depolarizing agents are known to increase cAMP levels.

In various species, treatment with NGF antiserum during the first few days after birth results in extensive destruction of pre- and paravertebral ganglia.[7,8] NGF therefore may be a prerequisite for the development of the peripheral sympathetic system, although other factors may also be responsible. Recent studies[21,25] in mice have shown that the formation of synapses between preganglionic cholinergic nerves and the cell body of the terminal adrenergic neurons precedes the development of TH activity. These observations imply that the development of the terminal adrenergic neurons depends not only on the effect of NGF but also on the activity of the preganglionic cholinergic nerve, or at least on the formation of synapses. However, it has also been found[26] that, in the 3-day-old rat, decentralization of the superior cervical ganglion delays but does not block general growth and differentiation of terminal adrenergic neurons. Two weeks after decentralization of the left superior cervical ganglion, CAT activity was determined as a measure of completeness of decentralization; protein content was determined as a measure of general growth; and the specific activities (product formed/hr/mg protein) of TH and DBH were determined as a measure of differentiation of terminal adrenergic neurons. CAT activity on the decentralized side was reduced to <3% of that on the intact side, the protein content to $68 \pm 3\%$, the specific activity of TH to $80 \pm 6\%$, and that of DBH to $79 \pm 8\%$. Decentralization did not impair the effect of NGF on general growth or the selective induction of TH and DBH. Therefore, it can be concluded that the main factor for growth and differentiation of the terminal adrenergic neuron is NGF. The action of NGF does not appear to depend on the intactness of the preganglionic cholinergic fibers, at least for the expression of the functions described above. The activity of preganglionic cholinergic fibers seems to represent only an additive factor in the development of the adrenergic neuron. Unlike the chick embryo, in mammals NGF causes hypertrophy of the sympathetic ganglia, because the dorsal ganglion cells appear to lose their sensitivity to NGF before birth.[27] NGF also causes the formation of aberrant sympathetic ganglion neuritic circuits to organs which normally do not receive adrenergic innervation.[28] The morphological and functional consequences of NGF-induced hypertrophy of mammalian dorsal root ganglion *in vivo* are difficult to study, possibly due to the problems involved in administrating NGF to the mammalian embryo *in vivo*. Crain and Peterson[29] employed an alternate approach to this problem, utilizing explants of fetal mouse spinal cord with attached dorsal root ganglion, which can develop organotypic structure and function during long-term culture. NGF causes hypertrophy of dorsal root ganglion in a 13- to 14-day-old fetal mouse spinal cord–dorsal root ganglion culture.[30] Correlative electrophysiological experiments were performed on these cultures to determine whether the NGF-stimulated growth and development of dorsal root ganglion neurons lead to enhancement of organized synaptic bundles and arborizations or are merely unregulated growth processes leading to "dead end" or aberrant functional connections. Crain and Peterson[30] have shown that at least some of the aberrant additional neurites that develop after exposure of fetal dorsal root ganglion cells to NGF make characteristic long-term synaptic relationships with specific types of spinal-cord neurons. It has also been determined[30] that responses resembling "primary afferent depolarization" can be generated in organized spinal cord–dorsal root ganglion explants, with remarkable mimicry of specialized network functions *in vivo*. The analysis of primary afferent depolarization generated in spinal-cord explants, in response to the unusually large input from NGF-hypertrophied dorsal root ganglia, shows their relationship to presynaptic inhibitory functions.[30] This may suggest the existence of an intrinsic central nervous system regulatory system, involving development of compensatory (homeostatic) inhibitory circuits proportional to the magnitude of the excitatory synaptic input.[31]

## Nerve Growth Factor and Serum Requirement

Dorsal root and sympathetic chick embryo ganglia exhibit a concentration-dependent increase in neurite outgrowth in medium with or without serum. However, there are some qualitative and quantitative differences under these two growth conditions. For example, no significant outgrowth of fibers occurs in the absence of nerve growth factor if the cultures are grown in serum-free medium, whereas a significant increase in neurite outgrowth occurs if the culture is grown in serum-supplemented medium. The minimum concentration of nerve growth factor required to achieve a significant increase in neurite outgrowth is higher in serum-free medium (18 ng/ml, NGF) than in serum-supplemented medium (0.4 ng/ml, NGF). The lag period between the addition of NGF and this response is much higher in ganglia grown in serum-free medium than that in serum-supplemented medium. These data show that serum factors may play a secondary or supporting role in mediating the effects of NGF, and/or that a portion of the neuronal population is relatively insensitive to NGF. The latter possibility is substantiated by the observation[32] that the ability of fetal calf serum to promote neurite outgrowth in dorsal root ganglion cultured in the absence of NGF is not inactivated by an antiserum directed against the $\beta$ subunit of 7 S NGF. In serum-supplemented medium the dorsal root ganglion of 8-day-old chick embryo was relatively more responsive than that of 12-day-old embryo.[32] This may reflect either a decrease in the sensitivity of 12-day-old dorsal root ganglion neurons to NGF or a decrease in the number of viable NGF-sensitive neurons, since there is more *in vivo* neurite outgrowth in the chick embryo DRG at 12 days than at 8 days of development.

### Nerve Growth Factor and cAMP

Recently, several laboratories have studied the question of whether or not cAMP is a mediator of the NGF response. Dibutyryl cAMP increases neurite formation in embryonic chick dorsal root ganglia[33] and mouse sensory root ganglia,[34] and elevates TH activity in cultures of sympathetic ganglia[35] and DBH activity in isolated superior cervical ganglia.[36] These changes are similar to those produced by NGF.[1-3,17,18] Based on evidence that both dibutyryl cAMP and NGF produce similar effects in chick embryo dorsal root ganglia on the criteria of growth responses, ultrastructure, and assembly of microtubules, it has been suggested[33,37] that NGF-induced neurite formation may be mediated by cAMP. On the other hand, other investigators have suggested that NGF and dibutyryl cAMP induce neurite formation by different mechanisms.[38,39] For example, dibutyryl cAMP does not increase the neurotubular protein content of dorsal root ganglia, whereas NGF increases this parameter by about twofold. This increase in neurotubular protein content after treatment with NGF can occur in the absence of neurite extension.[38] Several authors have concluded that the elevation in cellular

neurotubule subunit levels after NGF treatment may precede neurite
extension. However, this may be a secondary event unrelated to the
mechanism involved in NGF-induced neurite formation. This is not
sufficient evidence to rule out the involvement of cAMP in NGF-
induced responses. The fact that NGF neither increases the level of
cAMP in intact embryonic sensory ganglia, nor stimulates adenylate
cyclase activity in broken ganglia cells, suggests that NGF effects are
not mediated via cAMP.[139] The morphological effects of suboptimal
amounts of NGF on 9-day-old chick embryo sensory ganglion are not
enhanced by the addition of 0.5 mM theophylline. However, since
theophylline, in many intact systems, does not increase cAMP level,
the use of theophylline in the intact embryonic sensory ganglia may
not have provided conclusive data with respect to changes in the level
of cAMP after treatment with NGF. A recent study[40] has shown that
NGF (2.5 S) produces a severalfold increase in the cAMP content of
rat superior cervical ganglion in organ culture. This increase occurs
within 5 min of NGF addition and is followed by a return to the basal
level in about 10 min (Figure 4-2). An increase can be seen with as
little as 40 ng/ml of NGF (Figure 4-3). Oxidized NGF has no effect,

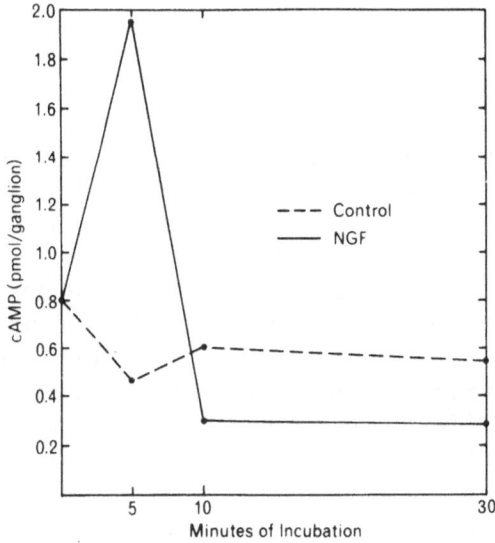

**Figure 4-2.** Temporal effect of NGF in rat superior cervical ganglia (60-min preincuba-
tion). 2.5 S NGF was added at a concentration of 3 $\mu$g/ml. A comparable amount of the
appropriate buffer was added in the control experiments. Mean ± S.E.M. (From
Nikodijevic *et al.*[40])

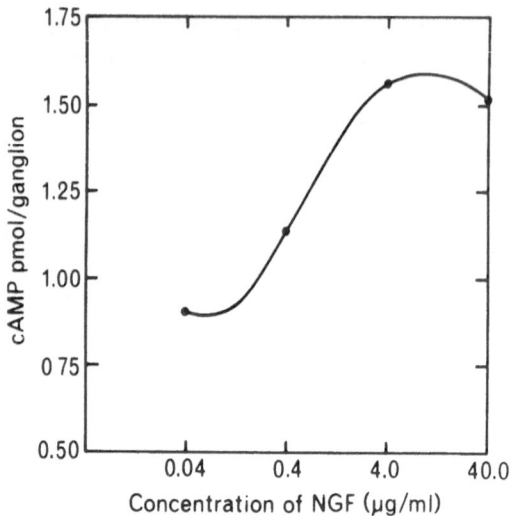

**Figure 4-3.** Effect of NGF concentration on cAMP level. Rat superior cervical ganglia were preincubated for 60 min at 37°. Then 2.5 S NGF was added, and the ganglia were incubated for an additional 5 min. Control value for cAMP was $0.68 \pm 0.03$ pmol/mg protein. Mean $\pm$ S.E.M. (From Nikodijevic *et al.*[40])

and increase in cAMP concentration produced by NGF is prevented by the addition of NGF antiserum.

### Nerve Growth Factor and Malignancy

NGF induces neurite outgrowth in certain lines of human neuroblastoma cells in culture,[41–43] human,[44,45] and rat[46] pheochromocytoma cells in culture, and in explants of human fetal adrenal medulla[47]; however, it fails to do so in mouse neuroblastoma cells.[43,48] The binding of NGF to the surface membrane of mouse neuroblastoma cells occurs at a higher level in $G_1$ and early S phases than in other phases of the cell cycle.[49] Sympathetic nerve cells bind NGF to their membrane surfaces to the same extent as mouse neuroblastoma cells.[50] Nevertheless, NGF fails to induce neurite formation in mouse neuroblastoma cells. It is possible that NGF, bound in these cells, fails to activate processes which organize the assembly of microtubules, a step necessary for the expression of neurites. NGF increases the level of microtubular subunit protein in mouse neuroblastoma cells[48] and in chick embryo dorsal root ganglion[38]; however, it fails to increase the level of 14-3-2, a neuronal-specific protein,[48] in human neuroblastoma cells. These data suggest that microtubular protein and 14-3-2 protein in neuroblastoma cells are independently regulated.[48]

Neither normal nor neoplastic pheochromocytes have long processes *in vivo;* therefore, the observation that NGF can induce process formation in these cells in culture[44-47] is of particular interest. These data suggest two possibilities: (1) The neoplastic pheochromotocytes have regained the neuronal feature, namely, responsiveness to NGF; or (2) all pheochromocyte cells have the capacity to form long processes, but they cannot express it due to the presence of inhibitory substances *in vivo.* Several pieces of evidence now indicate that the expression of neurite outgrowth and biochemically differentiated functions in neuronal cells in culture are independently regulated.[51] In addition, Lempinen[52] has suggested that glucocorticoids promote differentiation of fetal sympathoblasts into catecholamine-storing cells without processes. Neither hydrocortisone nor corticosterone affect NGF-induced process formation in rat pheochromocytoma cells in culture, indicating that factors other than glucocorticoids account for the absence of process formation *in vivo.*[46]

### Secretion of a Nerve Growth Factor

Several cell lines in culture secrete a biologically active nerve growth factor immunologically similar to mouse submaxillary gland NGF. These lines include mouse L cells, 3T3 cells, SV3T3 cells (3T3 cells transformed by simian virus),[53] primary chick embryo fibroblasts,[54] mouse neuroblastoma cells,[55] human glioblastoma,[56] rat glioma,[57] and primary human skin and synovial fibroblasts.[58] The significance of NGF secretion by these various cell lines is unknown. It has been suggested that secreted NGF may be responsible for the growth-promoting effect of conditioned tissue culture medium.[53,54] A given cell type may synthesize NGF, secrete it, bind it, and functionally respond to it as a part of an autoregulatory mechanism.[55] Although mouse neuroblastoma cells secrete NGF[55] and bind it to their membrane surfaces,[49] they do not form neurites after treatment with exogenous NGF (Prasad, unpublished observation). On the other hand, human neuroblastoma cells do show neurite outgrowth after treatment with NGF,[41,42] but it has not yet been demonstrated that they secrete NGF. NGF is found in human serum, but there is no significant difference between values obtained from patients with neuroblastoma in various stages of active disease, from patients free of disease for 3 years or more, from patients with tumors other than neuroblastoma, or from normal children.[59] Nerve growth-stimulating factor of human serum was similar to NGF.[59]

## Chemical Properties of Nerve Growth Factor

NGF from mouse salivary glands exists in at least two molecular-weight classes, depending upon the method of purification used. One of these classes is referred to as 7 S NGF (referring to its sedimentation coefficient), the other as 2.5 S NGF. The 7 S NGF is a complex of three different noncovalently bound proteins designated as $\alpha$, $\beta$, and $\gamma$.[60] Among these, only the $\beta$ component shows the characteristic biological activity of NGF in producing neurite outgrowth *in vitro*. Although the function of the 7 S NGF is not known, it may be related to storage of NGF within the submandibular gland.[61] If the biologically active subunit of NGF is isolated from pure 7 S complex (following pH-induced dissociation), the protein is referred to as $\beta$-NGF. However, if the same subunit is isolated from submandibular gland homogenate (without first purifying 7 S NGF), then the protein is referred to as 2.5 S NGF.[62] This species closely resembles $\beta$-NGF; it differs only in that limited proteolysis occurs at both amino and carboxy termini during isolation. These chemical differences apparently do not affect the biological activity of 2.5 S NGF.

It has been recently shown[63] that a solution of NGF at neutral pH comprises a monomer $\rightleftharpoons$ dimer equilibrium system, and at concentrations of protein usually employed in bioassay (1–10 ng/ml), the equilibrium mixture consists almost entirely of monomer. This suggests that the biological activity of native NGF at concentrations in the ng/ml range may be mediated by its monomeric form. A similar suggestion was made earlier.[64] However, the preparation of NGF cross-linked with dimethylsuberimidate produces biological effects on sensory ganglia similar to those produced by native NGF.[65] This argues against the notion that dimer NGF may be inactive.

NGF is also present in relatively high levels in certain snake venoms,[66] and the partial sequence of the factor obtained from the venom of the cobra *Naja naja* has been determined.[67] Like mouse NGF, snake NGF is also composed of two identical polypeptide chains, but with only 116 residues per chain. Each chain of mouse NGF contains 118 amino acids, 3 disulfide bonds, and a high amide content.[68-71] The sequence positions of some 70% of the residues have been established, and of these, the percentage of residues occupying identical sequence loci is about 65%, including the positions of six half cystines, suggesting an identical disulfide structure. These studies indicate that the primary structure of NGF has been appreciably conserved over a wide range of the evolutionary scale.[61]

The amino acid sequence of mouse NGF shows remarkable similarity to the primary structures of insulin and proinsulin.[72,73] For example, three of the six half-cystinyl residues which occur in both insulin and NGF are in identical positions, and two of these residues are paired in the same way in the two proteins. The sequence of mouse NGF can be matched with human proinsulin with only five deletions required to yield the maximum similarities of 21% identical residues. Moreover, the majority of identical residue positions are clustered in the segments of NGF that align with the A and B chain sections of proinsulin, separated by exactly the 35 residues required to accommodate the C peptide of proinsulin. Both insulin[74] and NGF appear to act on cell surfaces, and both elicit pleiotropic cellular metabolic events. These data suggest that NGF and insulin are related in the evolutionary sense and may have arisen from a common ancestral gene.[61] In spite of extensive work during the last three decades, the molecular mechanism of NGF action on nerve cells remains unknown. Many biological observations remain to be explained. For example, why do so many different cell types, both normal and malignant, secrete NGF, at least *in vitro*? Do these cells require NGF for their survival and/or growth, or do they secrete NGF for no biological purpose of their own? Does NGF always act alone on a cell, or may it also modify the effect of other hormones *in vivo*? What is the significance of NGF in human serum? Possibly the growth and differentiation of varieties of cells are controlled by proteins structurally similar to NGF. Slight variations in their structure may not affect their gross biological effects, but may be important in finer control of cell division and differentiation.

## EFFECT OF THYROID HORMONE

The effect of thyroid hormone on the neuronal maturation of the nervous system is well known and has been reviewed by Eayrs.[75] Thyroid tissue implanted in the fourth ventricle stimulates early maturation of the lid closure reflex in tadpole.[76] Thyroxine has a specific action on Mauthner's neurons in amphibians, causing them to regress during metamorphosis. Local application of thyroxine to the medulla of frog tadpoles results in premature regression of Mauthner's neurons and an increase in size of the surrounding neurons.[77] The mechanism of these thyroid hormone effects on neuronal maturation is unknown.

Speculations about the role of thyroid hormone in mammalian nerve cell differentiation have been primarily influenced by the hypothesis that in mammals there may be a critical period of development during which rapid changes in the morphology and physiology of prospective nerve and glial cells take place, analogous to changes occurring during amphibian metamorphosis. The notion that the thyroid gland influences neural development dates back centuries, to the observation that in mentally retarded human cretinoid infants the thyroid fails to develop normally. Cretinism is characterized by various degrees of mental retardation. The mental disturbance in hypothyroid infants becomes noticeable within 6 months after birth. The symptoms of cretinism can be reversed if thyroid hormone treatment is begun at an early age.[78] Neurological and mental disturbances are the most important lesions in severe congenital hypothyroidism.

Behavioral defects have been noted in cretinoid rats.[79,80] Present evidence suggests that thyroid hormone is neither available to nor required by the rat fetus during uterine life.[81] Therefore, the influence of thyroid hormone must be exerted during the postnatal period. There is a critical period of CNS development, during which changes in the level of thyroid hormone can cause either transitory acceleration or delay of some aspects of neural maturation. In the rat, this critical period extends over the first 10–14 days of life, and thereafter neither thyroidectomy nor excess thyroxine can change the course of normal development. Furthermore, thyroid medication, administered daily to the neonatally thyroidectomized rat, cannot reverse all of the damage to the maturing brain caused by the deprivation of thyroid hormone during the above critical period.[81] Studies on the structure and architecture of the brain of hypothyroid rats have revealed a reduced rate of growth of both perikarya and axons, resulting in a cortex in which the cell population appears more packed, the density of the axonal neuropil is diminished, and the basal dendrites associated with each neuron are shorter and less branched.[82,83] The development of the axonal network is particularly impaired in the region occupied by specific afferent plexuses (layer 4). It was suggested[84] that the interference with axonal and dendritic branching results in a drastic reduction in the probability of axodendritic interactions, which could be responsible for the behavioral deficits seen in the cretinoid rat. Hamburgh[85] observed that in the cerebral cortex of neonatally thyroidectomized rats the activities of succinic dehydrogenase and glutamate decarboxylase, which are associated with synaptosomes, mitochondria, and nerve terminals, are decreased; while the activity of lactate dehydrogenase, an enzyme associated with cell sap, and of glutamate dehydrogenase, an enzyme associated with mitochondria of nerve cell perikarya, were less affected by thyroid deprivation.

The formation of myelin sheath around the axon is one feature characteristic of neural differentiation. However, evidence concerning the effects of thyroid hormones on myelinogenesis is somewhat contradictory. Thyroxine consistently accelerated the appearance of myelin in thyroxin-supplemented cultures containing cerebella of newborn rats.[86] In some studies, a more rapid disappearance of newly formed myelin was observed in cerebellar cultures maintained in thyroxine-supplemented medium, as compared with those maintained without hormone. However, other workers[87] have found that thyroxine injected into immature rats during the period of active myelination had no significance on the ratio of myelin to nonmyelin phospholipids in the whole brain, although there was an increase in the total amount of brain phospholipid. Thyroxine may exert its effect on myelinogenesis by a one-step triggering reaction rather than by influencing a sequence of synthetic reactions of myelin precursors.[86] Thyroxine accelerates metabolic maturation of brain tissue, as measured by rates of $O_2$ consumption of glucose by brain slices.[88-90] Hamburgh et al.[92] reported an increase in $O_2$ consumption of cerebral cortex and cerebellum, but not of medulla oblongata, that was significant only during the first week of postnatal life. The period during which the maturing brain responds with an increased respiratory rate to the availability of excessive thyroxine is probably very limited. This might explain why thyroxine administration to the newborn animal did not cause the expected increase in $O_2$ consumption of the brain.[92] The very limited effect that thyroid withdrawal or excess thyroxine exerts on the respiratory metabolism of the developing nervous system contrasts with the well-known metabolic actions exerted by thyroxine on various adult tissues, and makes it unlikely that energy metabolism is the key mechanism through which this hormone influences growth and differentiation of nerve cells and nerve cell circuits.[91] It was demonstrated[93] that in the cerebral cortex the development of succinic dehydrogenase (Figure 4-4) and cholinesterase was irreversibly impaired when thyroid hormone was withheld from the newborn rat, but such treatment was without effect on the increase of aldolase and cytochrome oxidase in the maturing rat brain. Hormone therapy in thyroidectomized young rats, when started on

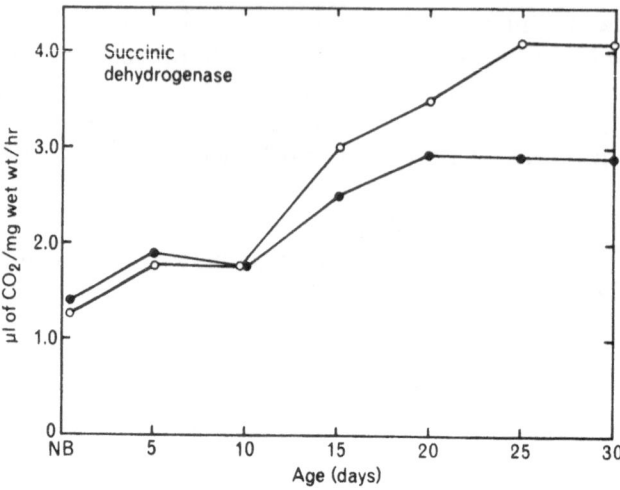

**Figure 4-4.** Each point represents the mean value of several determinations obtained from different litters for succinic dehydrogenase activity of cerebral cortex. NB, Newborn, ○, control rats; ●, thyroidectomized rats. (From Hamburgh and Flexner.[92])

the 10th day of postnatal life, led to a normal level of succinic dehydrogenase activity in young adults (25 days of age), but such therapy was without effect when the hormone was given at the 15th day or later. These authors suggested that in the developing cerebral cortex the deprivation of thyroid hormone leads to a decreased rate of synthesis of some selected protein, rather than to an overall lowering of various enzyme activities in this tissue. This interpretation was based on the observation that the decrease in enzyme activity coincided with the period of rapid protein increase in the brain of normal litter mates. In the rat brain numerous enzyme activities begin to increase after the 10th day, and continue to increase until adult levels are reached at day 25. Another observation in favor of their interpretation is the existence of a lag period between hormone manipulation and the observed end effect. The efficacy of thyroid replacement therapy in bringing enzyme activities to normal levels depends on whether therapy is instituted before or after the critical period of enzyme synthesis. This implies that, in the absence of thyroid hormone during this critical period, some irreversible changes may occur in the developing cerebral cortex. The role of thyroid hormone in the regulation of succinic dehydrogenase activity is further suggested by the observation that the activity of succinate dehydrogenase in cerebral cortex of hypothyroid decreases.[93,94] Glutamate dehydrogenase activity in the brain is less affected by thyroidectomy than is succinic dehydrogenase activity.[85] Thyroid hormone also seems to affect the activity of glutamate decarboxylase.[85] It has been reported[95] that thyroid hormone deficiency induced at birth by radioactive treatment is accompanied by a 15% decrease in total cholinesterase and a 26% decrease in AChE activity in cerebral cortex and hypothalamus. When enzyme activity was expressed per unit of DNA or per cell, the reduction in cholinesterase and AChE was 26% and 35%, respectively. The administration of thyroxine prevented these changes. Similarly, it was found[92] that there was an increase in AChE activity in cerebral cortex and hypothalamus in 13-day-old rats that had received exogenous thyroxine on postnatal days 2, 3, 4, and 5. The activities of these enzymes in the medulla oblongata and cerebellum were unaffected by hormone treatment.

The stimulatory effects of thyroxine on certain brain enzymes are paralleled by the effects of this hormone on overall protein synthesis, as measured by amino acid incorporation.[96] It was reported[97] that, in cell-free preparations from infant brain, thyroxine stimulated the amino acid incorporation into protein when hydroxybutyrate is the oxidizable substrate. However, this effect of the hormone was also demonstrated in the cell-free preparation from adult brain. In order to determine which of the cell fractions were involved, mitochondria, microsomes, and cell sap were isolated from both mature and immature brains. Geel et al.[98] observed that thyroxine stimulation of amino acid incorporation depended upon the source of mitochondrial fraction that was present in the incubation system. Replacement of the adult brain mitochondrial fraction by infant brain mitochondria resulted in thyroxine stimulation of amino acid incorporation by a cell-free system that consisted of amino acid incorporation by a cell-free system that consisted otherwise of components obtained from adult brain tissue. Replacement of the infant brain mitochondria with adult brain mitochondria in an otherwise "infant system" abolished thyroxine stimulation.[98] It was also observed[99, 100] that the specific activity of cerebral protein-bound leucine was reduced in hypothyroid rats following the injection of uniformly labeled [$^{14}$C]-L-leucine in proportion to body weight, indicating a decrease in protein synthesis. This deficiency in protein synthesis could be prevented by thyroxine medication. It is not clear if the effect of thyroid hormone occurs at the level of transcription or translation.

## Theories of Mechanism of Action of Thyroid Hormone

It has been postulated[91, 101] that thyroxine may push cells into the differentiative phase by taking them out of the proliferative or mitotic phase. By acting as a "timer," the optimal amount of thyroxine may turn off the proliferative phase and thus allow the differentiative phase to begin. Excess thyroxine may turn off the proliferative phase too soon, while thyroid deficiency may prolong it, thus delaying the onset of differentiation. Incorrect timing may therefore lead to both permanent as well as transitory abnormalities.

Another possibility is that thyroxine influences mammalian CNS differentiation by stimulating degenerative changes. This suggestion is based on the observation that, in the developing lateral motor column of *Rana pipiens* larvae, many of the cells present at the outset of development do not survive beyond the midlarval period.[102,103] A resorption effect of thyroxine was also demonstrated in the mammalian nervous tissue.[86,101] Explants of cerebella from newborn rats and mice maintained in culture continued differentiation into neurons and myelinated axons *in vitro*.[104] Addition of thyroxine to the culture medium hastened the breakdown of those myelin sheaths that had already formed, resulting in rapid disappearance of myelin. In unsupplemented control medium, myelin persisted for some time, even if the medium was not replaced for as long as 7–9 days. These data, as well as the demonstration by Weiss and Rossetti[77] that M cells atrophy under the influence of thyroid hormone, warrants further testing of the hypothesis that thyroxine may influence differentiation by two mechanisms, namely, the "phasing of proliferation" and the resorption of excess of cells.

## EFFECT OF GROWTH HORMONE

Growth hormone administered to pregnant rats results in a remarkable neuronal hyperplasia in the fetuses. The brain of the fetus shows a 10–20% increase in weight, DNA content, cortical cell density, and number and length of cortical dendrites, and in the ratio of neurons to glia[105–107] Growth hormone is effective only during the period of proliferation of matrix cells, which in the rat brain continues into the early neonatal period.[109,110] The functional result of the increase in the number of neurons, and probably also in the number of neuronal connections, is not entirely clear. The offspring of rats treated with growth hormone during pregnancy showed a 40% increase in the retention of a conditioned response compared with a control group of normal rats.[110] It has been assumed[107] that the increase in cell number is due to an extension of the proliferation period, and that the cells most sensitive to the hormone are neuroblasts, which continue to proliferate after birth. The consequences of an increased population of neuroblasts in the brain would not become apparent in the rat until after birth, when cell proliferation in the rat cortex has ceased and fiber outgrowth has begun.

The main objection to a direct effect of growth hormone on cerebral maturation during intrauterine life is the uncertainty that maternal pituitary hormone can cross the placenta.[111,112] The divergent reports on the modifications of cerebral development in mammalian embryos by growth hormone administration suggest that either these effects of the hormones are indirect, nonspecific, and mediated through influences on the mother, or that a small amount of the hormones can cross the placenta to the fetus. There is evidence to support both views. Excess growth hormone administered to the pregnant female might merely add to the maternal supply of nutrients, and protein deprivation before and during pregnancy can indeed result in a significant decrease in neonatal brain weight and brain DNA.[113] In a comparative study between two groups of pregnant rats, one maintained throughout gestation on an 8% protein-containing diet, the other on a 27% protein diet, it was found that the brain weight of newborn offspring was reduced by 23% and total brain DNA by 10% in the low-protein group. On the other hand, injection of 300 mg of bovine growth hormone into chick embryos increased neonatal brain DNA, and it seems unlikely that the addition of 300 mg of this hormone protein would significantly enrich the large nutritional depot in the chicken egg. Another possibility may be that growth hormone may influence brain maturation through a mechanism of action shared with other hormones. Biological systems can sometimes utilize alternative pathways to assure "important end results."[91,101] Indeed, it has been reported[114] that rats thyroidectomized at birth and treated daily with growth hormone recover their normal weight and their righting reflex. Furthermore, $O_2$ consumption of brain tissue, which was severely depressed by thyroid deprivation during postnatal maturation, returned to normal levels following growth hormone treatment. It has also been found that the impairment of thermoregulatory mechanisms typical of hypothyroid rats could be partially reversed by growth hormone replacement therapy. Hypothyroid rats given this hormone from birth were able to maintain an almost normal (35°C) internal body temperature upon reaching maturity, while untreated hypothyroid rats were unable to adjust their internal temperature above 28–30°C. These data suggest that growth hormone in the postnatal rat may function merely as one of several control mechanisms serving to diminish the need for thyroxine in subsequent development, and thus limit the critical period of thyroid sensitivity.

## CELL-CYCLE CHANGES DURING NEURONAL DIFFERENTIATION

A number of adult structures, including pigment cells of the skin, sensory and autonomic neurons, chromaffin cells of the adrenal medulla, some glial elements, and portions of the cranial skeleton, arise from neural crest cells in the embryo.[115-119] Prior to their differentiation, neural crest cells undergo a period of extensive migration from their origin in the neural folds and dorsal neural tube to their ultimate sites of localization. During and after their migratory phase, the neural crest cells also undergo extensive proliferation.[120-123] The precise control of cell migration and proliferation is essential for the normal development and morphogenesis of neural crest derivatives. Some chick neural crest cells *in vivo* differentiate into sensory ganglion neurons as early as 12 hr after the formation of the neural crest.[121] This neuronal differentiation involves extensive modification of the mitotic behavior of cells. It has been shown[122] that the rate of proliferation declines in the developing sensory ganglion of the chick between the 3rd and 7th day of development. Maxwell[123] has investigated changes in the cell cycle during neural crest differentiation *in vitro* of chick trunk neural tubes containing neural crest cells. Cell outgrowth from these neural tube explants consists primarily of a small stellate cell population. After 3 days in culture, the small stellate cell population undergoes a remarkable change in morphology, characterized by a more refractile appearance of cells in the phase-contrast microscope. Pigment granules become visible in the cytoplasm after 4 days in culture, and after 6 days, virtually all of the small stellate cells are pigmented. The cycle time $(G_1 + G_2 + M)$ increases from 7 hr on day 4 to 12 hr on day 5. This change is consistent with an increase in $G_1$ and/or $G_2$ that is closely correlated with the appearance of melanin granules. It should be noted that changes in the cell cycle are late events in the developmental program of these neural crest cells.[123]

## ROLE OF CELL INTERACTIONS IN THE REGULATION OF CHOLINERGIC ENZYMES DURING NEURAL DIFFERENTIATION

Cell interactions might play a role in the differentiation of various types of neural and glial cells from their common precursor cell, the neuroepithelial cell.[124-127] Adler *et al*[127] have used the 7-day-old chick embryo to investigate the role of cell interactions on CAT and AChE activities during differentiation. The neural retina, telencephalon, optic lobe, and rombencephalon were dissociated, and the resulting cell suspensions were allowed to reaggregate *in vitro* for 3 days, either independently or in different binary combinations. The combined neural retina–optic lobe aggregates showed an increase in CAT activity without changes in AChE activity; neural retina–telencephalon aggregates an increase in CAT and a decrease in AChE; telencephalon-rombencephalon aggregates a decrease in both CAT and AChE; and optic lobe–telecephalon aggregates no changes at all. These interactions require the presence of a precise number of cells of each kind.[128] When the ratio of neural retina and optic lobe cells is 90:10, no significant change in CAT activity is detected, in comparison to that observed in the individual culture; however, if the ratio is reversed, i.e., neural retina 10:optic lobe 90, a marked increase in CAT activity (105%) without any change in AChE occurs. A similar change in CAT activity (127%) occurs when the ratio of neural retina and optic lobe is 50:50. It remains to be established whether the cells of such an interacting unit have to be linked by permanent or transient functions, or whether the effect is mediated through the release of diffusible molecules. The interaction capability of a cell changes as a function of its developmental stages. The combined aggregates of neural retina–optic lobe from 10-day-old embryos did not significantly change CAT or AChE activity. This suggests that 10-day-old neural retina and optic lobe have completely lost the interaction properties they had at the 7-day stage. However, the combined aggregates of 10-day-old neural retina and 7-day-old optic lobe show a significant increase in both CAT and AChE activities, indicating that 10-day-old neural retina can interact with 7-day-old optic lobe cells. However, the effect of such interactions is somewhat different from that observed in the combined aggregates of neural retina–optic lobes from 7-day-old embryo, in which no change in AChE activity occurs. Thus, the interactive properties of 10-day-old neural retinal cells are somewhat different from those of 7-day-old neural retinal cells. Several possibilities can be suggested[128] to explain the mechanism of change in enzyme activity: (1) a differential survival of CAT- and/or AChE-containing cells; (2) a modification of these enzyme activities in cells already containing CAT and AChE molecules, either by activation or inhibition of preexisting molecules or by regulation of the synthesis of new ones; and (3) a stimulus for the differentiation of still uncommitted cells, in either a cholinergic or a noncholinergic direction. Although results obtained *in vitro* cannot readily be extrapolated to *in vivo* conditions, Adler *et al.*[127] have made some speculations about the possible roles that these interactions could play in normal development *in vivo*. It is well known that each cell normally develops within a changing cellular environment created by the temporospatial distribution of locally proliferating and differentiating cells, plus nerve fibers arriving from other neural organs. In addition, the activity of cholinergic enzymes depends, at least in the chick embryo optic lobe[129,130] and spinal cord,[131] upon the arrival of appropriate afferents. These regulatory processes could control the differentiation of neuroblasts as cholinergic or noncholinergic. Similar interactions could be important at earlier stages of development, since several authors have postulated that different neurotransmitters, including ACh, may have some morphogenetic and trophic functions in addition to their well-known role in synaptic transmission (review 132).

In the culture of neural crest cells, only melanocytes can be identified.[133,134] The apparent absence of other neural-crest-derived phenotypes *in vitro* is consistent with experiments showing that interaction with somatic mesenchyme conditions, by previous exposure to ventral neural tube, is apparently necessary for neural crest cells to differentiate into sympathetic neurons.[135] The possibility also exists that other cell interactions, not yet found in culture, are required for the differentiation and maintenance of other neural crest phenotypes.

## REGULATION OF NEURAL TUBE CLOSURE

Neural tube closure can be interfered with by the treatment of explanted early chick embryo (4–7 stages) with concanavalin A,[136] dinitrophenol,[137] actinomycin D,[138] cytochalasin B,[139] 5-BrdU,[140] etc. Each of these agents may have a specific mode of action in the chick neurulae. The data on Con A suggest[136] that this agent blocks closure of the neural tube by inhibiting interkinetic nuclear migration, irrespective of the developmental stage at treatment. However, erythropoiesis, migration of pericardiac cells to form tubular heart, regression of primitive streak, and segmentation of the axial mesoderm, usually were unaffected. The interkinetic nuclear migration is known to occur in the cells forming chick neural tube.[141,142] It has been observed[136] that cells capable of utilizing [³H]thymidine are located in the outer zone of neuroepithelium; subsequently, the labeled cells migrate to the neurocoel to undergo mitosis, after which they return to the outer zone to start new DNA synthesis prior to the next division. In Con-A-treated embryos, such nuclear migration was strongly inhibited and nuclei were often found to divide throughout the neuroepithelium.

## EFFECT OF NONNEURAL ELEMENTS

The neuronal survival of chick embryo dorsal root ganglia and sympathetic ganglia in dissociated culture depends strictly on the availability of NGF in the culture medium.[143,144] It has been shown[145] that nonneuronal elements play a key role in the survival and fiber production ability of dissociated chick enbryo sympathetic ganglion cells. It has been further suggested[146] that neuronal attachment and fiber production would not occur in cultures of a purified neuronal fraction, even in the presence of nerve growth factor, unless non-neural elements were supplied to them.

## EFFECT OF HEAVY WATER (D₂O)

Deuterium ($D_2O$), at a concentration of up to 25%, accelerates growth and maturation, and in some cases causes the multiplication of differentiated neurons in the explant culture of nervous tissue.[147] Histologically, different regions of the neuraxis differ in their response to deuterium. Sympathetic ganglia are the most greatly stimulated. $D_2O$ accelerates and increases growth of neurons and favors their repeated subdivision as an abnormally large size is attained. The total neuronal population eventually increases twofold or threefold; in turn, the progeny-clusters of small, neuroblastlike cells enlarge and become multipolar. Electron micrographs show an unusual abundance of fibrous and granular elements in the nuclei of both neurons and supporting cells. Fibrillar bundles, not seen in untreated cells, are found in the cytoplasm of deuterium-treated neurons; in addition, mitochondria may be abnormally dense, and microtubules are unusually numerous.

In sensory ganglia, growth and multiplication of supporting cells are the most conspicuous responses to $D_2O$.[147] Neurons become unusually turgid, and some time lobate. They tend to deteriorate as they are abandoned by the activated satellite cells.

The response of hypothalamus to $D_2O$ involves both neurons and glia. A neuron (with satellite cells attached) has been followed through mitosis at 16 days *in vitro*.[147] Growth of cerebellar neurons, extension of neurites, migration of glial cells, and myelination are accelerated by the presence of 10% $D_2O$ during early culture stages. The controls later catch up, however. In the cerebral cortex, 10% $D_2O$ accelerates maturation, but some dedifferentiation later follows.

The outgrowing nerve fibers from deuterated ganglionic explants appear earlier and grow to greater lengths than those of the controls, but the thick, bushy halo fiber typical of NGF-treated ganglionic culture is not observed.

## EFFECT OF SERUM-FREE MEDIUM AND X IRRADIATION

Serum-free medium induces neurite formation in both mouse[148] and human neuroblastoma cells[144]; however, it increases the intracellular level of cAMP by about twofold only in mouse cells.[149,150] These neurites regress after the addition of fresh growth medium. AChE activity of mouse neuroblastoma cells in serum-free medium increases,[148,141] but the activities of TH[151,152] and CAT (Prasad, unpublished observation) do not change. DA, which stimulates adenylate cyclase activity in homogenates of malignant mouse cells, fails to stimulate the enzyme activity in cells differentiating in serum-free medium.[153] Serum may be necessary for the expression of differentiated functions other than neurite formation and increase AChE activity.

X irradiation induces neurites and causes cell death in both mouse[154] and human[149] neuroblastoma cells, without changing the intracellular level of cAMP.[150] These neurites are electrically excitable.[155] The activities of CAT, AChE, and COMT increase in X-irradiated cells, but the activity of TH does not change.[156] The lack of stimulatory effect of X irradiation on TH activity is not due to damage of the biochemical machinery for TH synthesis, because irradiated mouse neuroblastoma cells show increased TH activity after treatment with sodium butyrate or dibutyryl cAMP. DA stimulates adenylate cyclase activity in malignant cells, but it does not do so in X-irradiated cells. Thus, X irradiation may produce some, but not all, of the molecular changes promoted by cAMP.[156]

## EFFECT OF 6-THIOGUANINE, CYTOSINE ARABINOSIDE, METHOTREXATE, AND 5-BROMODEOXYURIDINE

6-TG,[157] cytosine arabinoside,[151] methotrexate,[158] and 5-BrdU[159] induce neurite formation in mouse neuroblastoma cells. The effects of these agents vary from one clone to another. 6-Thioguanine increases the activities of CAT and COMT, but has no effect on TH activity.[157] Cytosine arabinoside increases AChE activity but has no effect on TH activity.[151] 6-TG does not increase cAMP levels.[157] The effect of cytosine arabinoside on cAMP levels has not been investigated. The effect of 6-TG on human neuroblastoma cells is less pronounced than on mouse neuroblastoma cells. 5-BrdU induces neurite formation and increases the activity of TH in both human and mouse neuroblastoma cells.[159,160] It doubles the cAMP level in mouse neuroblastoma cells, but has no such effect on human cells.[150,160]

## EFFECT OF GLIAL EXTRACT

An extract from glial cells induces neurite formation in mouse neuroblastoma cells without changing the growth rate of the cell population.[161] Extracts from other mammalian cell cultures, such as those of kidney cells from newborn hamster, fetal human fibroblasts, fetal mouse and rat glia, also produce similar morphological changes, but to a lesser degree. The intracellular level of cAMP is not significantly altered by glial extract, and it has therefore been suggested that cAMP may not be involved in differentiation. It has been shown that neurites can be formed in the absence of any change in the intracellular level of cAMP, but such morphologically differentiated cells do not express many of the biochemical functions characteristic of mature neurons. It remains to be seen whether neuroblastoma cells treated with glial extract can express any biochemically differentiated function.

## EFFECT OF HYPERTONIC MEDIUM

When mouse neuroblastoma cells are grown in monolayer culture in a medium made hypertonic with sucrose or sodium chloride (60% increase in osmolarity), cell division is inhibited almost immediately, and the cell number in the culture remains constant for up to 9 days. The neurites of these cells in hypertonic medium become exceedingly long and differentiated.[162] If at any time the hypertonic medium is replaced by normal medium, the neurites are retracted within 3 hr and cell division begins again. Whether the hypertonic medium induces biochemically differentiated functions is, again, not yet known.

## EFFECT OF COLLAGEN

Neurite outgrowth in cultured mouse neuroblastoma cells is reduced when the cells are grown on a collagen substratum rather than on glass. Rapid and synchronized initiation of neurite outgrowth occurs after hydrolysis of the underlying collagen with collagenase.[163] The cells with neurites exhibit an increase in RNA and protein contents compared to cells grown on collagen. This is in contrast to explants of spinal cord and spinal ganglia, which grow best on a collagen-coated surface.[164] Human neuroblastoma cells form neurites when grown on collagen-coated slides.[165] The inhibition of neurite formation in mouse neuroblastoma cells on collagen surfaces is thus surprising.

## EFFECT OF LIPOSOME

It has been reported[166] that liposome induces morphological differentiation of murine neuroblastoma; the mechanism of this effect is unknown.

## CONCLUSION

Nerve growth factor mimics many effects on nerve cells that are produced by cAMP. At least in rat superior cervical ganglia, NGF increases the cellular cAMP by severalfold within 5 min of NGF addition and this level returns to basal level 10 min after treatment. The exact relationship between nerve growth factor and cAMP remains to be established. Thyroid hormone also affects the differentiation of nerve cells. This is probably accomplished by taking cells out of a proliferative or mitotic phase and pushing them into a differentiatve phase, and by the resorption of excess cells. The administration of growth hormone to pregnant rats causes neuronal hyperplasia in fetuses. The mechanism of this effect is not well understood. In addition, it is not clear whether one hormone modifies the response of other hormones on developing cells. Although cell-cycle changes occur during differentiation, they are late events in the development program of nerve cells. Cell interaction plays an important role in the regulation of differentiated functions in nerve cells. It has been shown that, in addition to nerve growth factor, nonneural elements play a key role in the survival and fiber-production ability of the dissociated chick embryo sympathetic ganglia. Other agents (heavy water, serum-free medium, X irradiation, 6-TG, cytosine arabinoside, methotrexate, 5-BrdU, glial extract, hypertonic medium, liposome, etc.), are known to induce one or more differentiated functions in nerve cells or tissues in culture. The mechanisms of action of these agents are not well understood. However, some of these agents, such as serum-free medium (in mouse and human neuroblastoma cells) and 5-BrdU (only in mouse neuroblastoma cells) increase the intracellular level of cAMP by about twofold; however, X irradiation and 6-thioguanine do not change it. The effect of these agents on other parameters of cyclic nucleotide systems must be investigated before any conclusion can be made with respect to their mechanism of action.

## REFERENCES

1. Bueker, E. D., Implantation of tumors in the hind limb field of the embryonic chick and developmental response of the lumbosacral nervous system, *Anat. Rec.* **102**:369–390, 1948.
2. Levi-Montalcini, R., and Hamburger, V., A diffusible agent of mouse sarcoma producing hyperplasia of sympathetic ganglia and hyperneurotization of viscera in the chick embryo, *J. Exp. Zoo.* **123**:233–288, 1953.
3. Levi-Montalcini, R., and Angeletti, P. U., Growth control of the sympathetic system by a specific protein factor, *Q. Rev. Biol.* **36**:99–108, 1961.
4. Levi-Montalcini, R., and Angeletti, P. U., Essential role of the nerve growth factor in the survival and maintenance of dissociated sensory and sympathetic embryonic nerve cells *in vitro, Dev. Biol.* **7**:653–659, 1963.
5. Levi-Montalcini, R., Chemical stimulation of nerve growth, in: *Symposium on Chemical Basis of Development,* Vol. 234, pp. 646–664, Johns Hopkins University, McCollum-Pratt Institute, 1968.
6. Angeletti, P. U., Angeletti, R. H., Frazier, W. A., and Bradshaw, R. A., Nerve growth factor, in: *Proteins of the Nervous System* (D. J. Schneider, R. H. Angeletti, R. A. Bradshaw, A. Grasso, and B. W. Moore, eds.), pp. 133–154, Raven Press, New York, 1973.
7. Levi-Montalcini, R., and Booker, B., Destruction of the sympathetic ganglia in mammals by an antiserum to a nerve growth protein, *Proc. Natl. Acad. Sci. U.S.A.* **46**:384–391, 1960.
8. Levi-Montalcini, R. and Angeletti, P. U., Immunosympathectomy, *Pharmacol. Rev.* **18**:619–628, 1966.
9. Levi-Montalcini, R., Caramia, F., and Angeletti, P. U., Alterations in the fine structure of nucleoli in sympathetic neurons following NGF-antiserum treatment. *Brain Res.* **12**:54–73, 1969.
10. Angeletti, P. U., Luizzi, A., Levi-Montalcini, R., and Gandini-Attardi, D., Effect of a nerve growth factor on glucose metabolism by sympathetic and sensory nerve cells, *Biochim. Biophys. Acta* **90**:445–450, 1964.
11. Angeletti, P. U., Luizzi, A., and Levi-Montalcini, R., Stimulation of lipid biosyn-

thesis in sympathetic and sensory ganglia by a specific nerve growth factor, *Biochim. Biophys. Acta* **84**:778–781, 1964.

12. Angeletti, P. U., Gandini-Attardi, D., Toschi, G., Salvi, M. L., and Levi-Montalcini, R., Metabolic aspects of the effect of nerve growth factor on sympathetic and sensory ganglia: Protein and ribonucleic acid synthesis, *Biochim. Biophys. Acta* **95**:111–120, 1965.

13. Partlow, L. M., and Larrabee, M. G., Effects of a nerve growth factor, embryo age and metabolic inhibitors on growth of fibres and on synthesis of ribonucleic acid and protein in embryonic sympathetic ganglia, *J. Neurochem.* **18**:2101–2118, 1974.

14. Yamada, K. M., and Wessells, N. K., Axon elongation: Effect of nerve growth factor on microtubule protein, *Exp. Cell Res.* **66**:346–352, 1971.

15. Mizel, S. B., and Bamburg, J. R., Studies on the action of nerve growth factor. Role of RNA and protein synthesis in the process of neurite outgrowth, *Dev. Biol.* **49**:20–28, 1976.

16. Levi-Montalcini, R., The nerve growth factor: Its mode of action on sensory and sympathetic nerve cells, *Harvey Lect.* **60**:217–259, 1966.

17. Thoenen, H., Angeletti, P. U., Levi-Montalcini, R., and Kettler, R., Selective induction by nerve growth factor of tyrosine hydroxylase and dopamine $\beta$-hydroxylase in the rat superior cervical ganglia, *Proc. Natl. Acad. Sci. U.S.A.* **68**:1598–1602, 1971.

18. Stockel, K., Solomon, F., Paravicini, U., and Thoenen, H., Dissociation between effects of nerve growth factor on tyrosine hydrolase and tubulin synthesis in sympathetic ganglia, *Nature (London)* **250**:150–151, 1974.

19. Thoenen, H., Saner, A., Angeletti, P. U., and Levi-Montalcini, R., Increased activity of choline acetyltransferase in sympathetic ganglia after prolonged administration of nerve growth factor, *Nature (London), New Biol.* **236**:26–28, 1972.

20. Phillipson, O. T., and Sandler, M., The influence of nerve growth factor, potassium depolarization and dibutyryl (cyclic) adenosine 3′,5′-monophosphate on explant cultures of chick embryo sympathetic ganglia, *Brain Res.* **90**:273–281, 1975.

21. Black, I. B., Hendry, I. A., and Iversen, K. L., Trans-synaptic regulation of growth and development of adrenergic neurons in a mouse sympathetic ganglion, *Brain Res.* **34**:229–240, 1971.

22. Black, I. B., Hendry, I. A., and Iversen, L. L., Effect of surgical decentralization and nerve growth factor on the maturation of adrenergic neurons in a mouse sympathetic ganglion, *J. Neurochem.* **19**:1367–1377, 1972.

23. Hendry, I. A., and Iversen, L. L., Reducion in the concentration of nerve growth factor in mice after siaectomy and castration, *Nature (London)* **243**:500–504, 1973.

24. Stoeckel, K., Hendry, I. A., and Thoenen, H., Retrograde axonal transport of nerve growth factor, *Experientia* **29**:767, 1973.

25. Black, I. B., Bloom, F. E., Hendry, I. A., and Iversen, L. L., Growth and development of sympathetic ganglion: Maturation of transmitter enzymes and synapse formation in the mouse superior cervical ganglion, *J. Physiol. (London)* **215**:24–25, 1971.

26. Thoenen, H., Saner, A., Kettler, R., and Angeletti, P. U., Nerve growth factor and preganglionic cholinergic nerves; their relative importance to the development of the terminal adrenergic neuron, *Brain Res.* **44**:593–602, 1972.

27. Levi-Montalcini, R., and Booker, B., Excessive growth of the sympathetic ganglia evoked by a protein isolated from mouse salivary glands, *Proc. Natl. Acad. Sci. U.S.A.* **46**:373–384, 1960.

28. Olson, L., Outgrowth of sympathetic adrenergic neurons in mice treated with a nerve-growth factor (NGF), *Z. Zellforsch. Mikrosk. Anat.* **81**:155–173, 1967.

29. Crain, S. M., and Peterson, E. R., Onset and development of functional inter-neuronal connections in explants of rat spinal cord ganglia during maturation in culture, *Brain Res.* **6**:750–762, 1967.
30. Crain, S. M., and Peterson, E. R., Enhanced afferent synaptic functions in fetal mouse spinal cord-sensory ganglion explants following NGF-induced ganglion hypertrophy, *Brain Res.* **79**:145–152, 1974.
31. Crain, S. M., Tissue culture models of development of behavior and the nervous system, in: *Aspects of Neurogenesis*, Vol. 2 (G. Gottlieb, ed.), pp. 69–44, Academic Press, New York, 1974.
32. Mizel, S. B., and Bamburg, J. R., Studies on the action of nerve growth factor. 1. Characterization of simplified *in vitro* culture system for dorsal root and sympa-thetic ganglia, *Dev. Biol.* **49**:11–19, 1976.
33. Roisen, F. J., Murphy, R. A., Pichichero, M. E., and Braden, W. G., Cyclic adenosine monophosphate stimulation of axonal elongation, *Science* **175**:73–74, 1972.
34. Haas, D. C., Hier, D. B., Arnason, B. G. W., and Young, M., On a possible relationship of cyclic AMP to the mechanism of action of nerve growth factor, *Proc. Soc. Exp. Biol. Med.* **140**:45–47, 1972.
35. MacKay, A. V. P., and Iversen, L. L., Increased tyrosine hydroxylase activity of sympathetic ganglia cultured in the presence of dibutyryl cyclic AMP, *Brain Res.* **48**:424–426, 1972.
36. Keen, P., and McLean, W. G., Effect of dibutyryl cyclic AMP on levels of dopamine $\beta$-hydroxylase in isolated superior cervical ganglia, *Arch. Pharmacol.* **275**:465–469, 1972.
37. Roisen, F. J., and Murphy, R. A., Neurite development *in vitro*. II. The role of microfilaments and microtubules in dibutyryl adenosine 3′,5′-cyclic monophosphate and nerve growth factor stimulated maturation, *J. Neurobiol.* **4**:397–412, 1973.
38. Hier, D. B., Arnason, B. G. W., and Young, M., Studies on the mechanism of action of nerve growth factor, *Proc. Nat. Acad. Sci. U.S.A.* **69**:2268–2272, 1972.
39. Hier, D. B., Arnason, B. G. W., and Young, M., Nerve growth factor: Relation-ship to the cyclic AMP system of sensory ganglia, *Science* **182**:79–81, 1973.
40. Nikodijevic, B., Nikodijevic, O., Yu, M. W., Pollard, H., and Guroff, G., The effect of nerve growth factor on cyclic AMP levels in superior cervical ganglia of the rat, *Proc. Nat. Acad. Sci. U.S.A.* **72**:4769–4771, 1975.
41. Goldstein, M. N., Land, V., and Bradshaw, R., Stimulation of human neuroblas-tomas *in vitro* with nerve growth factor, *Proc. Am. Assoc. Cancer Res.* **13**:89, 1972.
42. Waris, T., Rechardt, L., and Waris, P., Differentiation of neuroblastoma cells induced by nerve growth factor *in vitro*, *Experientia* **29**:1128–1129, 1973.
43. Reynolds, C. P., and Perez-Polo, J. R., Human neuroblastoma: Glial induced morphological differentiation, *Neurosci. Lett.* **1**:91–97, 1975.
44. Tischler, A. S., Dichter, M., Biales, B., and Posner, M., The neural properties of pheochromocytoma cells in culture—Preliminary observations, *Lab. Invest.* **32**:437–438, 1975.
45. Kadin, M. E., and Bensch, K. G., Comparison of pheocromocytes with ganglion cells and neuroblasts grown *in vitro*, *Cancer* **27**:1148–1160, 1971.
46. Tischler, A. S., and Greene, L. A., Nerve growth factor-induced process forma-tion by cultured rat pheochromocytoma cells, *Nature (London)* **258**:341–342, 1975.
47. Hervonen, A., and Kanerva, L., Neuronal differentiation in human fetal adrenal medulla, *Int. J. Neurosci.* **5**:43–46, 1973.
48. Kolber, A. R., Goldstein, M. N., and Moore, B. W., Effect of nerve growth factor

on the expression of colchicine-binding activity and 14-3-2-protein in an established line of human neuroblastoma, *Proc. Nat. Acad. Sci. U.S.A.* **71**:4203–4207, 1974.

49. Revoltella, R., Bertolini, L., and Pediconi, M., Unmasking of nerve growth factor membrane-specific binding sites in synchronized murine C1300 neuroblastoma cells, *Exp. Cell Res.* **85**:89–94, 1974.

50. Revoltella, R., Bosman, C., and Bertolini, L., Detection of nerve growth factor binding sites on neuroblastoma cells by rosette formation, *Cancer Res.* **35**:890–895, 1975.

51. Prasad, K. N., Differentiation of neuroblastoma cells in culture, *Biol. Rev.* **50**:129–165, 1975.

52. Lempinen, M., Extra-adrenal chromaffin tissue of the rat and the effect of cortical hormones on it, *Acta Physiol. Scand., Suppl. 231* **62**:7–91, 1964.

53. Oger, J., Arnason, B. G. W., Pantazis, N., Lehrich, J., and Young, M., Synthesis of nerve growth factor by L and 3T3 cells in culture, *Proc. Natl. Acad. Sci. U.S.A.* **71**:1554–1558, 1974.

54. Young, M., Oger, J., Blanchard, M. H., Asdourian, H., Amos, H., and Arnason, B. G. W., Secretion of a nerve growth factor by primary chick fibroblast cultures, *Science* **187**:361–362, 1975.

55. Murphy, R. A., Pantazis, N. J., Arnason, B. G. W., and Young, M., Secretion of a nerve growth factor by mouse neuroblastoma cells in culture, *Proc. Natl. Acad. Sci. U.S.A.* **72**:1895–1898, 1975.

56. Arnason, B. G. W., Oger, J., Pantazis, N. J., and Young, M., Secretion of nerve growth factor by cancer cells, *J. Clin. Invest.* **53**:2a, 1974.

57. Longo, A. M., and Penhoet, E. E., Nerve growth factor in rat glioma cells, *Proc. Natl. Acad. Sci. U.S.A.* **71**:2347–2349, 1974.

58. Saide, J. D., Murphy, R. A., Canfield, R. E., Skinner, J., Robinson, D. R., Arnason, B. G. W., and Young, M., Nerve growth factor in human serum and its secretion by human cells in culture, *J. Cell Biol.* **67**(Suppl. 376a), 1975.

59. Bill, A. H., Seibert, E. S., Beckwith, J. B., Hartmann, J. R., Nerve growth factor and nerve growth stimulating activity in sera from normal and neuroblastoma patients, *J. Natl. Cancer Inst.* **43**:1221–1230, 1969.

60. Varon, S., Nomura, J., and Shooter, E. M., Reversible dissociation of the mouse nerve growth factor protein into different subunits, *Biochemistry* **7**:1296–1303, 1968.

61. Bradshaw, R. A., and Young, M., Nerve growth factor—Recent developments and perspectives, *Biochem. Pharmacol.* **25**:1445–1449, 1976.

62. Bocchini, V., and Angeletti, P. U., The nerve growth factor. Purification of a 30,000-molecular-weight protein, *Proc. Natl. Acad. Sci. U.S.A.* **64**:787–794, 1969.

63. Young, M., Saide, J. D., Murphy, R. A., and Arnason, B. G. W., Molecular size of nerve growth factor in dilute solution, *J. Biol. Chem.* **251**:459–464, 1976.

64. Frazier, W. A., Boyd, L. F., and Bradshaw, R. A., Interaction of nerve growth factor with surface membranes: Biological competence of insolubilized nerve growth factor, *Proc. Natl. Acad. Sci. U.S.A.* **70**:2931–2935, 1973.

65. Stach, R. W., and Shooter, E. M., The biological activity of cross-linked β nerve growth factor protein, *J. Biol. Chem.* **249**:6668–6674, 1974.

66. Cohen, S., and Levi-Montalcini, R., A nerve growth-stimulating factor isolated from snake venom, *Proc. Natl. Acad. Sci. U.S.A.* **42**:571–577, 1956.

67. Hogue-Angeletti, R. A., Frazier, W. A., Jacobs, J. W., Niall, H. D., and Bradshaw, R. A., Purification, characterization, and partial amino acid sequence of nerve growth factor from cobra venom, *Biochemistry* **15**:26–34, 1976.

68. Angeletti, R. H., Hermodson, M. A., and Bradshaw, R. A., Amino acid sequences of mature 2.5S nerve growth factor. II. Isolation and characterization of the thermolytic and peptic peptides and the complete covalent structure, *Biochemistry* **12**:100–115, 1973.

69. Angeletti, R. H., Bradshaw, R. A., and Wade, R. D., Subunit structure and amino acid composition of mouse submaxillary gland nerve growth factor, *Biochemistry* **10**:463–469, 1971.

70. Greene, L. A., Varon, S., Piltch, A., and Shooter, E. M., Substructure of the β-subunit of mouse 7S nerve growth factor, *Neurobiology* **1**:37–48, 1971.

71. Bocchini, V., The nerve growth factor amino acid composition and physic chemical properties, *Euro. J. Biochem.* **15**:127–131, 1970.

72. Frazier, W. A., Angeletti, R. H., and Bradshaw, R. A., Nerve growth factor and insulin, *Science* **176**:482–488, 1972.

73. Frazier, W. A., Hogue-Angeletti, R. A., Sherman, R., and Bradshaw, R. A., Topography of mouse 2.5S nerve growth factor reactivity of tyrosine and tryptophan, *Biochemistry* **12**:3281–3293, 1973.

74. Cuatrecasas, P., Interaction of insulin with the cells membrane: The primary action of insulin, *Proc. Natl. Acad. Sci. U.S.A.* **63**:450–457, 1969.

75. Eayrs, J. T., Influence of the thyroid on the central nervous system, *Brit. Med. Bull.* **16**:122–127, 1960.

76. Kollros, J. J., Localized maturation of lid-closure reflex mechanism by thyroid implants into tadpole hind-brain, *Proc. Soc. Exp. Biol. Med.* **49**:204–206, 1942.

77. Weiss, P., and Rossetti, F., Growth responses of opposite sign among different neuron types exposed to thyroid hormones, *Proc. Natl. Acad. Sci. U.S.A.* **37**:540–556, 1951.

78. Brown, A. W., Bronstein, I. P., and Kraines, R., Hypothyroidism and cretinism in childhood. VI. Influence of thyroid therapy on mental growth, *Am. J. Dis. Child.* **57**:517–523, 1939.

79. Eayrs, J. T., and Horn, G., The development of cerebral cortex in hypothyroid and starved rats, *Anat. Rec.* **121**:53–61, 1955.

80. Hamburgh, M., Sobel, E. H., Koblin, R., and Rinestone, A., Passage of thyroid hormone across the placenta in intact and hypophysectomized rats, *Anat. Rec.* **144**:219–227, 1962.

81. Eayrs, J. T., and Taylor, S. H., The effect of thyroid deficiency induced by methyl thiouracil on the maturation of the central nervous system, *J. Anat.* **85**:350–358, 1951.

82. Eayrs, J. T., The cerebral cortex of normal and hypothyroid rats, *Acta Anat.* **25**:160–183, 1955.

83. Eayrs, J. T., Thyroid and central nervous development, in: *Scientific Basis of Medicine Annual Reviews* (J. P. Ross, ed.), pp. 317–339, Athlone Press, London, 1966.

84. Balazs, R., Kovács, S., Teichgraber, P., Cocks, W. A., and Eayrs, J. T., Biochemical effects of thyroid deficiency on the developing brain, *J. Neurochem.* **15**:1335–1349, 1968.

85. Hamburgh, M., Evidence for a direct effect of temperature and thyroid hormone on myelinogenesis *in vitro*, *Dev. Biol.* **13**:15–30, 1966.

86. Myani, N. B., and Cole, L. A., Effect of thyroxine on the deposition of phospholipids in the brain *in vivo* and on the synthesis of phospholipids by brain slices, *J. Neurochem.* **13**:1299–1307, 1966.

87. Reiss, J. M., Reiss, M., and Wyatt, A., Action of thyroid hormones on brain metabolism of newborn rats, *Proc. Soc. Exp. Biol. Med.* **93**:19–22, 1956.

88. Hoexter, F. M., The effect of thyroidectomy on the $O_2$ uptake of brain cortex and liver in the rat, *Endocrinology* **54**:1–4, 1954.

89. Fazekas, J. F., Graves, F. B., and Alman, R. W., The influence of the thyroid on cerebral metabolism, *Endocrinology* **48**:169–174, 1951.

90. Hamburgh, M., Lynn, E., and Weiss, E. P., Analysis of the influence of thyroid hormone on prenatal and postnatal maturation of the rat, *Anat. Rec.* **150**:147–162, 1964.

91. Schapiro, S., Some physiological, biochemical and behavioral consequences of neonatal hormone administration: Cortisol and thyroxine, *Gen. Comp. Endocrinol.* **10**:214–228, 1968.

92. Hamburgh, M., and Flexner, L. B., Biochemical and physiological differentiation during morphogenesis. XXI. Effect of hypothyroidism and hormone therapy on enzyme activities of the developing cerebral cortex of the rat, *J. Neurochem.* **1**:279–288, 1957.

93. Garcia Argiz, C. A., Pasquini, J. M., Kaplún, B., and Gomez, C. J., Hormonal regulation of brain development. II. Effect of neonatal thyroidectomy on succinate dehydrogenase and other enzymes in developing cerebral cortex and cerebellum of the rat, *Brain Res.* **6**:635–646, 1967.

94. Geel, S. E., and Timiras, P. S., Influence of neonatal hypothyroidism and of thyroxine on the acetylcholinesterase and cholinesterase activities in the developing central nervous system of the rat, *Endocrinology* **80**:1069–1074, 1967.

95. Klee, C. B., and Sokoloff, L., Mitochondrial differences in mature and immature brain. Influence on rate of amino acid incorporation into protein and responses to thyroxine, *J. Neurochem.* **11**:709–716, 1964.

96. Gelber, S., Campbell, P. L., Deibler, G. E., and Sokoloff, L., Effects of L-thyroxine on amino acid incorporation into protein in mature and immature rat brain, *J. Neurochem.* **11**:221–229, 1964.

97. Sokoloff, L., Action of thyroid hormones and cerebral development, *Am. J. Dis. Child.* **114**:498–506, 1967.

98. Geel, S. E., Valcana, T., and Timiras, P. S., Effect of neonatal hypothyroidism and of thyroxine on L-[$^{14}$C]leucine incorporation in protein *in vivo* and the relationship to ionic levels in the developing brain of the rat, *Brain Res.* **4**:143–150, 1967.

99. Geel, S. E., and Timiras, P. S., The influence of neonatal hypothyroidism and of thyroxine on the ribonucleic acid and deoxyribonucleic acid concentrations of rat cerebral cortex, *Brain Res.* **4**:135–142, 1967.

100. Hamburgh, M., An analysis of the action of thyroid hormone on development based on *in vivo* and *in vitro* studies, *Gen. Comp. Endocrinol.* **10**:198–213, 1968.

101. Beaudoin, A. R., The development of lateral motor column cells in the lumbo-sacral cord in Rana pipiens, *Anat. Rec.* **121**:81–91, 1955.

102. Race, J., Jr., Thyroid hormone control of development of lateral motor column cells in the lumbo-sacral cord in hypophysectomized *Rana pipiens*, *Gen. Comp. Endocrinol.* **1**:322–331, 1961.

103. Bornstein, M. B., and Murray, M. R., Serial observation on patterns of growth, myelin formation, maintenance and degeneration in cultures of newborn rat and kitten cerebellum, *J. Biophys. Biochem. Cytol.* **4**:499–504, 1958.

104. Zamenhof, S., Stimulation of cortical-cell proliferation by the growth hormone. III. Experiments on albino rats, *Physiol. Zool.* **15**:281–292, 1942.

105. Clendinnen, B. G., and Eayrs, J. T., The anatomical and physiological effects of prenatally administered somatotrophin on cerebral development in rats, *J. Endocrinol.* **22**:183–193, 1961.

106. Zamenhof, S., Mosley, J., and Schuller, E., Stimulation of the proliferation of cortical neurons by prenatal treatment with growth hormone, *Science* **152**:1396–1397, 1966.
107. Altman, J., and Das, G. D., Post-natal origin of microneurones in the rat brain, *Nature (London)* **207**:953–956, 1965.
108. Altman, J., and Das, G. D., Autoradiographic and histological evidence of post-natal hippocampal neurogenesis in rats, *J. Comp. Neurol.* **124**:319–336, 1965.
109. Block, J. B., and Essman, W. B., Growth hormone administration during pregnancy: A behavioural difference in offspring of rats, *Nature (London)* **205**:1136–1137, 1965.
110. Gitlin, D., Kumate, J., and Morales, C., Metabolism and maternofetal transfer of human growth hormone in the pregnant woman at term, *J. Clin. Endocrinol.* **25**:1599–1608, 1965.
111. Laron, Z., Pertzelan, A., Mannheimer, S., Goldman, J.,and Guttmann, S., Lack of placental transfer of human growth hormone, *Acta Endocrinol.* **53**:687–692, 1966.
112. Zamenhof, S., Van Marthens, E., and Margolis, F. L., DNA (cell number) and protein in neonatal brain: Alteration by maternal dietary protein restriction, *Science* **160**:322–323, 1968.
113. Gomez, C. J., Ghittoni, N. E., and Dellacha, J. M., Effect of L-thyroxine or somatotrophin on body growth and cerebral development in neo-natal thyroidectomized rats, *Life Sci.* **5**:243–246, 1966.
114. Hörstadius, S., *The Neural Crest*, Oxford University Press, London, 1950.
115. Weston, J. A., The migration and differentiation of neural crest cells, *Adv. Morphog.* **8**:41–114, 1970.
116. Johnston, M. C., A radioautographic study of the migration and fate of cranial neural crest cells in the chick embryo, *Anat. Rec.* **156**:143–156, 1966.
117. LeDouarin, N. M., and Teillet, M. M., Experimental analysis of the migration and differentiation of neuroblasts of the autonomic nervous system and of neuroectodermal mesenchymal derivatives using a biological cell marking technique, *Dev. Biol.* **41**:162–184, 1974.
118. Noden, D. M., An analysis of migratory behavior of avian cephalic neural crest cells, *Dev. Biol.* **42**:106–130, 1975.
119. Weston, J. A., A radioautographic analysis of the migration and localization of trunk neural crest cells in the chick, *Dev. Biol.* **6**:279–310, 1963.
120. Hamburger, V., and Levi-Montalcini, R., Proliferation, differentiation, and degeneration in the spinal ganglia of the chick embryo under normal and experimental conditions, *J. Exp. Zool.* **111**:457–500, 1949.
121. Yates, R. D., A study of cell deivision in chick embryo ganglia, *J. Exp. Zool.* **147**:167–182, 1961.
122. Maxwell, G. D., Cell cycle changes during neural crest cell differentiation *in vitro*, *Dev. Biol.* **49**:66–79, 1976.
123. Weiss, P., Neurogenesis, in: *Analysis of Development* (B. H. Willier, P. Weiss, and V. Hamburger, eds.), 346 pp., Saunders, Philadelphia, 1955.
124. Sperry, R. W., Embryogenesis of behavioral nerve nets, in: *Organogenesis* (R. L. DeHaan and H. Ursprung, eds.), 161 pp., Holt, Rinehart and Winston, New York, 1965.
125. Alder, R., Cell interactions and histogenesis in embryonic neural aggregates, *Exp. Cell Res.* **77**:367–375, 1973.
126. Morris, J. E., and Moscona, A. A., The induction of glutamine synthetase in cell

aggregates of embryonic neural retina: Correlations with differentiation and multi-cellular organization, *Dev. Biol.* **25**:420–444, 1971.

127. Adler, R., Teitelman, G., and Suburo, A. M., Cell interactions and the regulation of cholinergic enzymes during neural differentiation *in vitro*, *Dev. Biol.* **50**:48–57, 1976.

128. Filogamo, G., Recherches expérimentales sur l'activité des cholinestérases spécifique et non spécifique dans le développement du lobe optique du poulet, *Arch. Biol.* **71**:159–198, 1960.

129. Marchisio, P. C., Choline acetyltransferase (ChAc) activity in developing chick optic centres and the effects of monolateral removal of retina at an early embryonic stage and at hatching, *J. Neurochem.* **16**:665–671, 1969.

130. Burt, A. M., and Narayanan, C. H., Effect of extrinsic neuronal connections on development of acetylcholinesterase and choline acetyltransferase activity in the ventral half of the chick spinal cord, *Exp. Neurol.* **29**:201–210, 1970.

131. Karczmar, A. G., Srinnivasan, R., and Bernsohn, J., Cholinergic functions in the developing fetus, in: *Fetal Pharmacology* (L. Boréus, ed.), pp. 127–177, Raven Press, New York, 1973.

132. Maxwell, G. D., Cell cyclic changes during neural crest cell differentiation *in vitro*, *Dev. Biol.* **49**:66–79, 1976.

133. Dorris, F., Differentiation of pigment cells in tissue cultures of chick neural crest, *Proc. Soc. Exp. Biol. Med.* **34**:448–449, 1936.

134. Cohen, A. M., Factors directing the expression of sympathetic nerve traits in cells of neural crest origin, *J. Exp. Zool.* **179**:167–182, 1972.

135. Norr, S. C., *In vitro* analysis of sympathetic neuron differentiation from chick neural crest cells, *Dev. Biol.* **34**:16–38, 1973.

136. Lee, H., Inhibition of neurulation and interkinetic nuclear migration by concanavalin A in explanted early chick embryos, *Dev. Biol.* **48**:392–399, 1976.

137. Bowman, P., The effect of 2,4-dinitrophenol on the development of early chick embryos, *J. Embryol. Exp. Morphol.* **17**:425–431, 1967.

138. Lee, H., and Poprycz, W., Effect of actinomycin D on explanted early chick embryos, *Growth* **34**:437–454, 1970.

139. Linville, P. G., and Shepard, T. H., Neural tube closure defects caused by cytochalasin B, *Nature (London) New Biol.* **236**:246–247, 1972.

140. Lee, H., Deshpande, A. K., and Kalmus, G. W., Studies on effects of 5-bromodeoxyuridine on the development of explanted early chick embryos, *J. Embryol. Exp. Morphol.* **32**:835–848, 1974.

141. Sauer, F. C., The interkinetic migration of embryonic epithelial nuclei, *J. Morphol.* **60**:1–11, 1936.

142. Watterson, R. L., Structure and mitotic behavior of the early neural tube, in: *Organogenesis* (R. L. DeHaan, and H. Ursprung, eds.), pp. 129–159, Holt, Rinehart and Winston, New York, 1965.

143. Levi-Montalcini, R., and Angeletti, P. U., Essential role of the nerve growth factor in survival and maintenance of dissociated sensory and sympathetic embryonic nerve cells *in vitro*, *Dev. Biol.* **7**:653–659, 1963.

144. Cohen, A. I., Nicol, E. C., and Richter, W., Nerve growth factor requirement for development of dissociated embronic sensory and sympathetic ganglia in culture, *Proc. Soc. Exp. Biol. Med.* **116**:784–789, 1964.

145. Varon, S., and Raiborn, C., Dissociation, fractionation and culture of chick embryo sympathetic ganglionic cells, *J. Neurocytol.* **1**:211–221, 1972.

146. Varon, S., Raiborn, C., and Tyszka, E., In vitro studies of dissociated cells from newborn mouse dorsal root ganglia, Brain Res. 54:51–63, 1973.
147. Murray, M. R., and Benitez, H. H., Action of heavy water (D₂O) on growth and development of isolated nervous tissue, in: Growth of the Nervous System (G. E. W. Wolstenholme and M. O'Connor, eds.), pp. 148–178, Little, Brown, 1968.
148. Seeds, N. W., Gilman, A. G., Amano, T., and Nirenberg, M. W., Regulation of axon formation by clonal lines of a neural tumor, Proc. Natl. Acad. Sci. U.S.A. 66:160–167, 1970.
149. Prasad, K. N., and Kumar, S., Role of cyclic AMP in differentiation of human neuroblastoma cells in culture, Cancer 36:1338–1343, 1975.
150. Prasad, K. N., Gilmer, K., and Kumar, S., Morphologically "differentiated" mouse neuroblastoma cells induced by non-cyclic AMP agents. Levels of cyclic AMP, nucleic acid and protein, Proc. Soc. Exp. Biol. Med. 143:1168–1171, 1973.
151. Kates, J. R., Winterton, R., and Schlesinger, K., Induction of acetylcholinesterase activity in mouse neuroblastoma tissue culture cells, Nature (London) 229:345–346, 1971.
152. Waymire, J. C., Weiner, N., and Prasad, K. N., Regulation of tyrosine hydroxylase activity in cultured mouse neuroblastoma cells. Elevation induced by analogs of adenosine 3',5'-cyclic monophosphate, Proc. Natl. Acad. Sci. U.S.A. 69:2241–2245, 1972.
153. Prasad, K. N., and Gilmer, K. N., Demonstration of dopamine-sensitive adenylate cyclase in malignant neuroblastoma cells and change in sensitivity of adenylate cyclase to catecholamines in "differentiated" cells, Proc. Natl. Acad. Sci. U.S.A. 71:2525–2529, 1974.
154. Prasad, K. N., X-ray-induced morphological differentiation of mouse neuroblastoma cells in vitro, Nature (London) 234:471–474, 1971.
155. Nelson, P., Ruffner, W., and Nirenberg, M., Neuronal tumor cells with excitable membranes growth in vitro, Proc. Natl. Acad. Sci. U.S.A. 64:104–110, 1969.
156. Prasad, K. N., Differentiation of neuroblastoma cells in culture, Biol. Rev. 50:129–165, 1975.
157. Prasad, K. N., Differentiation of neuroblastoma cells induced in culture by 6-thioguanine, Int. J. Cancer 12:631–635, 1973.
158. Byfield, J. E., and Karlsson, V., Inhibition of replication and differentiation in malignant mouse neuroblastoma, Cell Differ. 2:55–64, 1973.
159. Schubert, D., and Jacob, F., 5-Bromodeoxyuridine-induced differentiation of a neuroblastoma, Proc. Natl. Acad. Sci. U.S.A. 67:247–254, 1970.
160. Prasad, K. N., Mandal, B., and Kumar, S., Human neuroblastoma cell culture. Effect of 5-bromodeoxyuridine on morphological differentiation and levels of neural enzymes, Proc. Soc. Exp. Biol. Med. 144:38–42, 1973.
161. Monard, D., Solomon, F., Rentsch, M., and Gysin, R., Glial-induced morphological differentiation in neuroblastoma cells, Proc. Natl. Acad. Sci. U.S.A. 70:1894–1897, 1973.
162. Ross, J., Granett, S., and Rosenbaum, J. L., Differentiation of neuroblastoma cells in hypertonic medium, J. Cell Biol. 59:291a, 1973.
163. Miller, C. A., and Levine, E. M., Neuroblastoma synchronization of neurite outgrowth in cultures grown on collagen, Science 177:799–802, 1972.
164. Bornstein, M. B., Reconstituted rat-tail collagen used as substrate for tissue cultures on coverslips in Maximow slides and roller tubes, Lab. Invest. 7:134–137, 1958.

165. Burdman, J. A., and Goldstein, M. N., Long-term tissue culture of neuroblastoma. III. *In vitro* studies of a nerve-growth stimulating factor in sera of children with neuroblastoma, *J. Natl. Canc. Inst.* **33**:123–133, 1964.
166. Chen, J. S., Del Fa, A., DiLuzio, A., and Calissano, P., Liposome-induced morphological differentiation of murine neuroblastoma *Nature (London)* **263**:604–606, 1976.

# PARENT AND HYBRID NEUROBLASTOMA CELLS IN CULTURE AS A MODEL SYSTEM FOR NEURONAL FUNCTION

## INTRODUCTION

Parent and hybrid neuroblastoma cells lines have been extensively used to study certain aspects of neurobiology. *In vitro* hybridization of somatic cells of dissimilar differentiated types results in extinction of some differentiated phenotypes, continued expression of others, and even activation of some traits.[1] In order to understand the mechanism of regulation of differentiated functions in nerve cells, mouse neuroblastoma cells have been fused with three different cell types: fibroblasts, glioma cells, and sympathetic ganglion cells.

## HYBRID CELLS (MOUSE NEUROBLASTOMA CELLS × MOUSE FIBROBLAST CELLS)

Various neuronal features such as AChE, neuronal morphology, action potential generation, ACh sensitivity, neural-specific protein 14-3-2, and glycosphinogolipid, which are present in neuroblastoma cells, were expressed in the hybrid cells to a varying degree.[2-7] Steroid sulfatase was extinguished in all hybrid cells when only one parent expressed this enzyme, but was expressed in one hybrid combination in which both parents expressed this enzyme.[2] One neuronal enzyme was activated in hybrid cells derived from a parental clone that lacked this property.[3]

## HYBRID CELLS (MOUSE NEUROBLASTOMA CELLS × RAT GLIOMA CELLS)

Dibutyryl cAMP induces neurite formation in hybrids,[8] similar to that reported for parent neuroblastoma cells.[9] Although dibutyryl cAMP causes neurite extension in two hybrid cell lines, it increases DBH activity in only one of them. This means that in the hybrid cells the extension of processes is not necessarily coupled with the increase in DBH activity.[8] A similar conclusion has been reached for many biochemically differentiated functions in parent neuroblastoma cells.[10] Certain hybrids have CAT, which is not exhibited by either parental cell line.[11,12] The ultrastructural features of neurons present in the hybrid cells are markedly stimulated by dibutyryl cAMP.[13]

## HYBRID CELLS (MOUSE NEUROBLASTOMA CELLS × MOUSE SYMPATHETIC GANGLION CELLS)

The fusion of mouse neuroblastoma cells (N18 TG2), which lacks tyrosine 3-monooxidase activity, with cells from mouse sympathetic ganglia that have this enzyme yields a somatic cell hybrid line that has tyrosine 3-monooxidase activity and synthesizes dihydroxyphenylalanine (dopa) and DA.[12] The hybrid cells, after treatment with dibutyryl cAMP, resemble neurons in morphology and possess highly excitable membranes that are capable of generating action potentials. In addition, these cells have nicotinic and possibly muscarinic excitory ACh receptors.[14] Thus, the fusion of neuroblastoma cells with normal cells from the nervous system can yield hybrid cells with new neuronal properties. Such hybrid cell lines should be useful as a model system for studying synapse formation and other neural properties.

## HYBRID CELLS IN THE STUDY OF NARCOTIC ADDICTION

The presence of opiate receptors in some neuroblastoma-derived cell lines and in a hybrid cell line (neuroblastoma × glioma, NG108-15) has been reported.[15,16] Using the hybrids of neuroblastoma × glioma, it has been shown[15,17,18] that the binding of morphine and other opiates to narcotic receptors results in an inhibition of adenylate cyclase, and a decrease in the intracellular cAMP level.[15,17] Narcotics do not normally occur *in vivo;* therefore, opiate receptors must recognize endogenous ligands, which are mimicked by the opiates. The opiate agonist binding is favored at low concentrations of sodium ions and antagonist binding at high concentrations of sodium ions.[19,20]

Narcotics antagonize the effect of $PGE_1$ on neuroblastoma cells[15,17,21] as well as in brain homogenates.[22,23] It has been suggested[24] that narcotics affect adenylate cyclase in two ways, both mediated by the opiate receptor. The first process is a readily reversible inhibition of the enzyme by narcotics. The second phenomenon, late positive regulation, results in an increase in adenylate-cyclase-specific activity dependent upon incubation of cells with narcotics for 12 or more hours. The effects of narcotics on both inhibition and positive regulation of adenylate cyclase are stereospecific and are reversible with naloxone. After culture with morphine for 1–2 days, cells are tolerant to the narcotic, since the increase in adenylate cyclase activity due to positive regulation compensates for the inhibition of enzyme activity observed in the presence of narcotic. Cells are also dependent upon morphine, since withdrawal or displacement of morphine from the opiate receptor by the antagonist, naloxone, increases basal-, $PGE_1$-, and adenosine-stimulated adenylate cyclase activities and results in an overproduction of cAMP. (A model for the role of adenylate cyclase regulation in the development of morphine tolerance and dependence is shown in Figure 5-1.) The development of an increased response to $PGE_1$ under the influence of morphine can be completely blocked by cycloheximide. In animals, the development of narcotic tolerance can be prevented by inhibition of protein synthesis.[25]

The mechanism of increase in adenylate cyclase activity after prolonged exposure of neuroblastoma hybrids to narcotics is unknown. A few suggestions have been made. One group of investigators has suggested[24] that the positive regulation of adenylate cyclase represents a new type of receptor-mediated control of adenylate cyclase activity.[24] Inhibition of adenylate cyclase activity and late positive regulation are coupled, since both are initiated by interactions of narcotic with the opiate receptors. The inhibition of adenylate cyclase activity and the concomitant reduction in cAMP levels may be required for positive regulation; i.e., the concentration of cAMP or the activity of adenylate cyclase may regulate the number of adenylate cyclase molecules in a sequential feedback fashion. If so, each neurotransmitter, hormone, or effector of adenylate cyclase may in fact be a dual regulator for transient stimulation of inhibition of adenylate cy-

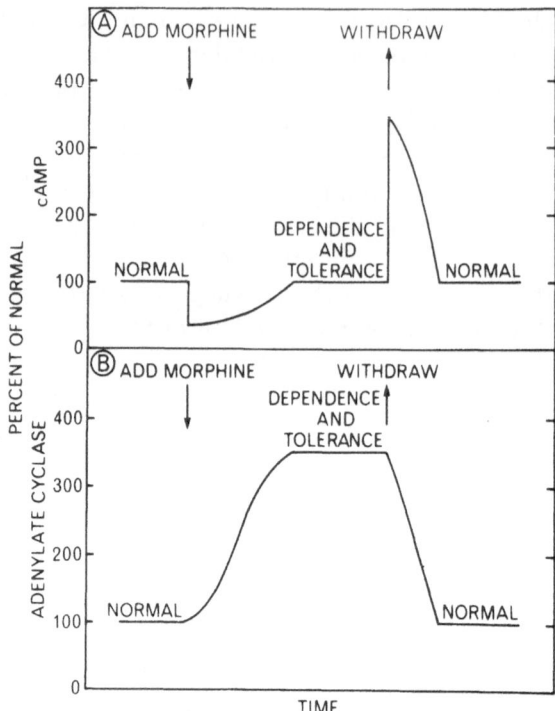

**Figure 5-1.** A model of the role of adenylate cyclase regulation in the development of morphine tolerance and dependence. Part A shows the effects of morphine upon cAMP levels; part B, the effect of the opiate upon adenylate cyclase activity as a function of time. (From Sharma *et al.*[24])

clase, and may envoke a regulatory process in the opposite direction hours after the initial event. The slowly expressed, relatively stable, positive regulatory process is a form of memory. Shifts in the number of molecules of adenylate cyclase or a long-lived modulator of enzyme activity would alter the sensitivity of cell response to other activators or inhibitors of adenylate cyclase and thus would markedly affect the efficiency of transsynaptic transmission. Morphine stimulates butyrylcholinesterase and AChE activities after prolonged treatment of human neuroblastoma cells (IMR-32 cell line).[26]

## UPTAKE STUDY USING PARENT NEUROBLASTOMA CLONES

### Choline Uptake

Choline probably cannot be synthesized by neurons[27-29]; therefore, the transport of choline across the neuronal membrane is essential for phospholipid metabolism and acetylcholine synthesis. Choline uptake might be rate-limiting in the synthesis of ACh.[30,31] The uptake of choline in serum-free-medium-induced differentiated cells was higher than that in control neuroblastoma cells.[32] The uptake in the inactive clone (N18), having no CAT activity but a barely detectable level of TH activity, was inhibited if sodium ions were substituted with cesium, lithium, potassium, or sucrose. Incubation with potassium ferricyanide or at a low temperature inhibited the incorporation of [$^{14}$C]choline in both control and serum-free-medium-induced differentiated cells. Thus, there is a strict relationship between choline uptake and the sodium pump, and an energy-dependent active component is involved in choline transport in this inactive clone. Two mechanisms, which differ in their affinity for choline, have been identified.[33] Clone S21 (CAT present; no detectable TH) showed a high-affinity component with an apparent $K_m$ that was much lower ($1.1 \times 10^{-7}$ M) than the one observed in clone N18 ($2.6 \times 10^{-6}$ M). This may be partly due to the cholinergic property of S21 clone. Choline uptake in cholinergic neuroblasts is markedly reduced by inhibitors of cholinesterase activity. The high-affinity $K_m$ was absent in a pure glial population obtained after dissociation of chick embryo cerebral hemisphere.[34] This implies that the high-affinity carrier for choline uptake is present only in the neuron, irrespective of the presence or absence of the cholinergic property. Removal of a portion of cell surface by sialic acid markedly reduced choline uptake and concurrently enhanced AChE activity.[35] With other clones, the high-affinity uptake of choline is absent in certain neuroblastoma cell lines.[36,37] A low-affinity uptake system, similar to that observed in nonneural cells, is present in these neuroblastoma cells. The reasons for this discrepancy among cell lines cannot be explained at this time.

## Norepinephrine Uptake

There are specialized transport systems, mediating NE uptake from the extracellular fluid in various types of neurons in peripheral and central noradrenergic nerve terminals. A specific and saturable high-affinity uptake for the NE precursor tyrosine has been demonstrated in the adrenergic clone NIE-115.[37] Two mechanisms of NE transport system were found in neuroblastoma adrenergic clone $M_1$.[38] One was a saturable high-affinity transport system that was partially energy-dependent and temperature-sensitive and required sodium, and could be inhibited by ouabain, amphetamine, imipramine, and desipramine. A second was a saturable low-affinity uptake that was also partially energy-dependent and temperature-sensitive, but was

not inhibited by the drugs tested. The uptake system of NE in neuroblastoma cells seems to have properties similar to those observed for
neuronal uptake.[39] Both mouse neuroblastoma (NB-2a) and rat C6
glioma showed ouabain-sensitive $K^+$ uptake, which correlated with
the level of $[Na^+ + K^+]$-ATPase activity found in the respective total
cell homogenates. The glioma cells had a 2.1-fold higher rate of $K^+$
uptake than neuroblastoma cells, and a 2.4-fold higher $[Na^+ + K^+]$-
ATPase activity.[40] In the presence of ouabain, neuroblastoma cells
released $K^+$ and took up $Na^{2+}$ in a 1:1 ratio. Such a mechanism is
known to exist in nervous tissue.

## Dopamine and 6-Hydroxydopamine Uptake

In monolayer culture of NIE-115 neuroblastoma cells, the high-
affinity uptake system for dopamine is lacking. Accumulation of
6-hydroxydopamine (6-OHDA) in cells of this clone was fourfold
higher than that of DA.[41] Reserpine inhibited the uptake of DA, but
not that of 6-OHDA. Furthermore, when uptake was followed by a
release period, cells retained a much greater proportion of the
6-OHDA radioactivity compared to that of the DA. However, radioactive 6-OHDA was more readily displaced by a higher concentration of
either 6-OHDA or DA than radioactive DA during the release period.
It has been suggested that the cytotoxicity of 6-OHDA and related
compounds in neuroblastoma cells[42,43] results from the ability of such
compounds to cross-link proteins.[41]

## γ-Aminobutyric Acid Uptake

It is well established that the principal mode of inactivation of
neurotransmitters other than ACh is by reuptake into presynaptic
terminals.[44] However, the close spatial relationship of neurons and
glia, and the glial ensheathment of axons and synapses, suggests that
glia, as well as neurons, may participate in the uptake of synaptically
released neurotransmitters.[45,46] Evidence that mammalian brain accumulates exogenous γ-aminobutyric acid (GABA) has been derived
from studies on tissue slices, synaptosomal preparations, and frac-
tionated cells. However, the cellular heterogeneity inherent in these
preparations has made it difficult to identify the cells involved in
GABA uptake. Clonal glioma (C6) and neuroblastoma (NB41) cell
lines take up GABA from the medium by a high-affinity, carrier-
mediated transport system ($K_m = 0.2$ μM) and a second, apparently
nonsaturable, process that may represent diffusion.[47] The high-
affinity transport of GABA in both cell lines is sodium dependent.

The similarity of GABA transport in these glial and neuroblastoma cell lines supports the hypothesis that glial cells may participate in the inactivation of GABA by reuptake from the synaptic cleft. Aminooxyacetic acid inhibits GABA uptake in C6 astrocytoma cells, but not in NB41 neuroblastoma cells. This difference suggests that GABA uptake in glia, but not in neurons, may be coupled to its metabolism.

Using clonal nerve cells and glial cells (derived from chemically induced brain tumor), it has been shown[48] that these cells are able to accumulate exogenous GABA. The ability of the cells to concentrate GABA is dependent upon external calcium, while the uptake process is sodium-dependent and temperature-sensitive. GABA uptake in cell lines is specific with respect to analog inhibition; this specificity is different from that observed in brain slices, but it is similar to that of glia in the rat sensory ganglia.

## INTRACELLULAR BIOGENIC AMINES

Some clones of mouse neuroblastoma cells contain catecholamines.[49,50] A recent study[51] using fluorescence-microscope histochemistry has demonstrated the presence of biogenic amines in an NB-2a clone of neuroblastoma and C6 glioma. Both cell lines synthesize monoamines. However, the neuroblastoma clone has two cell populations; one synthesizes primarily catecholamines and the other synthesizes primarily 5-hydroxytryptophan (serotonin). Recent studies indicate that nerve cells may synthesize and store more than one neurotransmitter at a time. We first demonstrated that a neuroblastoma clone NBP$_2$ contains extremely low activity of TH, but relatively high activity of CAT.[52] The low activity of TH can be stimulated 80-fold by dibutyryl cAMP. In a hybrid clone (neuroblastoma × glioma) the presence of DBH and CAT as well as the presence of ACh and catecholamines has been demonstrated.[13] Several neural cell lines (derived from CNS tumor) contain more than one neurotransmitter synthetic enzyme.[53] For example, B65 contains CAT, TH, and glutamic acid decarboxylase (GAD); B35 and B103 contain CAT and GAD. Of several proteins which seem to be localized in the nervous system, S-100 and 14-3-2 are thought to be associated with glia and nerve cells, respectively,[54,55] despite exceptions.[56-59] More recently, it has been found[53] that the presence of these proteins is not uniquely associated with either neuronal or nonneuronal cells. It has been recently reported[60] that in cultures of developing chick embryo nerve cells, both catecholamines and acetylcholine are present in the same neurons. Thus, at least tumor and developing nerve cells can synthesize and store more than one neurotransmitter at a time. It is unknown if such neurons exist in adult mammalian nervous tissue. The existence of such neurons would violate Dale's principle that one neuron cannot synthesize and store more than one neurotransmitter at a time.

## TRANSPORT OF GLUCOSE

Mouse neuroblastoma cells (NB41A$_3$) take up 2-deoxy-D-glucose (2-DG) and 3-O-methyl-D-glucose by a saturable transport process.[61] Experiments with various inhibitors (phloritin, pholorizin, cytochalasin B, ouabain, sodium cyanide, and iodoacetate) showed that this process is passive, sodium-dependent, and carrier-mediated. In rapidly growing culture (20 × 10$^3$ cells/cm$^2$), $K_m$ and $V_{max}$ for the transport of 2-DG were 0.8 mM and 18.3 nmol/min/mg protein, respectively. An increase in cell density (40 × 10$^3$ cells/cm$^2$) was followed by a drastic drop in the specific uptake, mainly due to a change in $V_{max}$. In stationary-phase culture (120 × 10$^3$ cells/cm$^2$), $K_m$ and $V_{max}$ for the transport of 2-DG were 4.9 mM and 9.1 nmol/min/mg protein, respectively. D-Glucose was a competitive inhibitor with a $K_i$ of 3.2 mM. Nonspecific uptake was 35–50% of total uptake at the cell densities studied. The temperature coefficient for the transport of 2-DG into both rapidly growing and stationary phase cultures was 1.4 (37°), while the activation energies differed, and were found to be 4.6 and 8.9 kcal/mol, respectively. In contrast to the transport of 2-DG in stationary phase, the uptake of [$^3$H]-2-DG in cAMP-induced differentiated cells[62] remained unchanged.

The transport of glucose into cultured human glioma cells (138-MG) as a function of cell density was in part different from that observed in mouse neuroblastoma.[63] The increase in 2-DG uptake in glioma cells during the first 3 days of culturing was dependent on the number of days in culture. The decrease in 2-DG uptake into glioma cells from day 3 to day 7 and into neuroblastoma cells throughout the experimental period was density-dependent. The inhibition of 2-DG uptake is not related to the inhibition of growth rate, since 2-DG uptake is reduced at cell densities at which no significant reduction in growth rate was observed. At the stationary phase when cell division was reduced, a marked increase in uptake of 2-DG occurred in glioma cells, but a marked decrease occurred in neuroblastoma cells. It has been reported[64] that glycogenolysis and subsequent outflow of glucose is activated by NE in cultured glioblastoma but not in neuroblastoma cells. Although the expression of certain differentiated functions of nerve cells is increased at confluency, many functions do not change or decrease. Therefore, the confluent culture of neuroblastoma cells may not be a suitable model to study the regulation of differentiated functions.

## METABOLISM OF GLYCOGEN

Using C6 glioma and NB-2a neuroblastoma cells, the regulation of glycogen metabolism was studied.[65] Many investigators have considered that glycogen is largely confined to glial elements, and particularly astrocytes,[66] although an analysis of single cells has shown that glycogen is present in at least some of the large neurons of the central nervous system.[67] There appear to be at least two modes of regulating glycogen metabolism in nerve and glial cells, namely, the available energy supply (glucose) and altered cAMP level. When glucose is added to starved cells, glycogen synthesis is turned on (glycogen phosphorylase $a$ decreases and glycogen synthetase $a$ increases). As glucose disappears from the medium, glycogenolysis is favored (phosphorylase $a$ increases and glycogen concentration decreases). The critical glucose concentration is 2.5 mM for glioma cells and 4 mM for neuroblastoma cells. Insulin promotes the conversion of phosphorylase $a$ to the $b$ form and synthetase $b$ to the $a$ form in both cell lines. All these changes occur without alterations in the intracellular level of cAMP. When cAMP concentrations are increased in either cell line, phosphorylase $a$ is increased, synthetase $a$ is decreased, and glycogen concentration is decreased. Isobutyl methylxanthine is effective in promoting glycogenolysis in both cell lines, whereas NE is effective in glioma cells, and $PGE_1$ is effective in neuroblastoma cells.

## DISTRIBUTION OF FREE NUCLEOTIDES

The pattern of free nucleotides in neuroblastoma cells (M and N115) differs from that in adult mouse brain tissue, where adenylic nucleotides represent 70% and uridylic nucleotides represent about 10% of total nucleotides.[68] The relative low percentage of adenylic nucleotides and the high amount of UPD-sugars in neuroblastoma cells is closer to the distribution of free nucleotides in very young brain tissue. The amounts of UDP-coenzymes progressively decrease while the amounts of adenylic nucleotides increase during postnatal development.[69] The distribution of free nucleotides in neuroblastoma cells suggests a potential for the synthesis of a large amount of UPD-coenzymes but a lower availability of high-energy adenylic nucleotides compared to adult brain. It is unknown if a switch from a lower to a higher availability of adenylic nucleotides occurs during differentiation of neuroblastoma cells.

## DISTRIBUTION OF FREE AMINO ACIDS

Concentrations of free amino acids ($\gamma$-aminobutyric acid, $\beta$-alanine, glutamic acid, aspartic acid, asparagine, glycine, and taurine) in several lines of nerve and glial cells (derived from chemically induced CNS tumor) were determined at the stationary phase of growth when cell division stops.[70] The relative concentrations of free amino acids described in the nerve and glial cell lines are in general agreement with those published for whole brain,[71] although there are exceptions. For example, the concentration of glutamic acid was the highest of the free amino acids in whole cerebral tissue,[71] and in the nerve and glial cells,[70] but the concentration of aspartic acid was significantly higher in the brain than in the nerve and glial cell lines.

## PHOSPHOLIPID METABOLISM

Wide varieties of chemical and physical stimuli, including neurotransmitters, hormones, and electrical impulses, stimulate the incorporation of $^{32}P_i$ into phosphatidylinositol and phosphatidic acid in secretory and nervous tissues,[72,73] a phenomenon also called the phospholipid effect. ACh is known to enhance the labeling of these acidic phospholipids in brain slices and synaptosomal preparations.[74-76] NE also evokes an increased turnover of acidic phospholipids in brain preparations,[77] including synaptosomes,[78] and in the rat pineal gland.[79,80] However, both carbamylcholine and NE failed to elicit a phospholipid effect[81] in clonal (NIE) neuroblastoma cells at stationary growth phase or in a clonal rat glial cell line ($C_{21}$) at confluent growth phase. Thus, neuroblastoma and glial cells in culture differ from adult nervous tissue on the criteria of ACh and catecholamine effects on phospholipid metabolism. It would be interesting to see if these cells would show a response similar to that of adult nervous tissue after treatment with cAMP-stimulating agents.

## IMMUNOLOGY OF NEUROBLASTOMA CELLS

At least five antigens have been described in mouse neuroblastoma.[82,83] In addition, tumor-specific antibodies may be eluted from neuroblastoma cells.[84] These eluted antibodies, when radiolabeled, absorbed with normal tissues, and injected into tumor-bearing mice, showed a highly specific uptake in neuroblastoma tumor. Pretreatment of syngenic strain A mice with aldehyde-fixed neuroblastoma cells (clone NB6R) almost completely protected the mice against challenge with viable NB6R cells.[85] In contrast, tumor growth was enhanced in mice treated with fixed cells after challenge with viable cells. Thus, depending on the time immunization was initiated, the procedure could be either protective or harmful. One study[86] has shown that murine neuroblastoma *in vivo* can be cured by an antibody-dependent cellular cytotoxicity reaction. The morphological differentiation of mouse neuroblastoma cells, induced by serum-free medium, results in an antigenic alteration in the cell membrane.[87] Cell-surface antigens are present that are unique to and/or highly enriched on morphologically differentiated cells. Brain particulate preparations contain antigens that can react with and remove antibodies directed to the antigens found on morphologically differentiated cells, while similar amounts of other organs' preparations are unable to absorb such antibodies.

Carlin *et al.*[88] reported that neurofilament protein isolated from calf brain is antigenically related to a component of the bundles of 100-Å filaments in neuroblastoma cells. This neurofilament protein (54,000–56,000 daltons) is an integral part of bundles of 100-Å filaments in neuroblastoma, while neither tubulin nor tropomyosin is present in these bundles.

## EFFECT OF QUINAZOLINE ANTIFOLATES

Correlations between inhibition of growth and of enzyme activity by several quinazoline analogues of folic acid were made using two lines of mouse neuroblastoma cells, one sensitive and one resistant to quinazoline analogues.[89] The 2,4-diamino-quinazoline analogues DAQ (N-[p-[(2,4-diamino-6-quinazolinyl)methyl]methylamino] benzoyl]- L- glutamic acid) and methasquin (N- [p- [(2,4- diamino- 5- methylquinazolinyl) methylamino] benzoyl]- L- aspartic acid) effectively inhibited cell growth as well as dihydrofolate reductase activity, whereas the 2- amino- 4- hydroxyquinazoline analogue AHQ (N- [p-

[[2-amino-4-hydroxy-6-quinazolinyl)methyl]methylamino]ben-
zoyl]-L-glutamic acid) was less potent. Nevertheless, AHQ inhibited
thymidylate synthetase activity more than DAQ. There was a good
correlation between inhibition of growth and dihydrofolate reductase. When 2,4-diaminoquinazoline analogues were used, however,
AHQ-induced inhibition depended upon the inhibition of both dihydrofolate reductase and thymidylate synthetase. Sensitive cells
could be completely protected against the toxic effect of AHQ by
either leucovorin or thymidine. Because of its unique potency towards thymidylate synthetase, AHQ may be useful in the
chemotherapy of neuroblastoma.[89]

## EFFECT OF NONNEURAL CELLS ON THE EXPRESSION OF
## NEURONAL FUNCTIONS

The importance of morphological and metabolic coupling between neurons and glia in the nervous system is well established.
However, the mechanisms involved in the interaction of these cells is
rather limited. It has been reported[90-94] that contact with glial cells
enhances differentiation of dissociated neurons, and maintains them
in long-term culture. ACh synthesis from radioactive choline is increased 100- to 1000-fold in the presence of nonneural cells from
sympathetic ganglia of newborn rat.[95] This increase was roughly dependent on the number of ganglionic nonneural cells present. This
effect was not due to an increased plating efficiency of neurons, since
the nonneural cells were capable of increasing ACh synthesis after
only a 48-hr contact with neurons that had been previously grown
without nonneural cells for 2 weeks. Rat glioma (C6) also stimulated
ACh synthesis, but mouse 3T3 cells had little or no effect.[95] One of
the nonneuronal cell types synthesized detectable ACh in the absence
of the neurons. The ganglionic nonneural cells had no significant
effect on catecholamine synthesis that occurs in the absence of nonneural cells.[95] It has been shown[96] that cocultivation of mouse
neuroblastoma and glioma cell lines markedly enhanced $Mg^{2+}$- and
$Ca^{2+}$-dependent ecto-ATPase activity. The increase in ecto-ATPase
activity of cocultured neuroblastoma and glioma cells occurs in both
cell types. Ecto-ATPase activity was 50% of the original level in clonal
line NN astroblasts after coculture with $M_1$ neuroblasts. This activity
decreased over 50 passages during the period of about a year. Increase in ecto-ATPase and morphological differentiation of $M_1$
neuroblastoma cells after coculture with NN astroblasts could be

brought about also by treatment with conditioned medium from NN cell culture. Cocultivation of neuroblastoma and glioma cells does not significantly change the specific activity of ecto-5'-nucleotidase.[96] CAT activity was more than 10-fold greater in combined culture of spinal cord and muscle cells than in cultures of spinal cord alone.[97] Therefore, it appears that the study of cocultivation of glial and neuronal culture may provide useful information with regard to the mechanisms by which one cell type modifies the expression of biological functions of the other.

## CONCLUSION

The parent and hybrid neuroblastoma cell lines have been useful in the study of many aspects of neurobiology, such as the molecular mechanism of narcotic addiction, uptake of neurotransmitters, metabolism of glycogen and phospholipids, distribution of free nucleotides and free amino acids, and the immunology of nerve cells. These cell lines have assisted in the study of the effect of nonneural cells on the expression of neuronal functions. This aspect of study has just begun to emerge. The results obtained from neuroblastoma study have raised an important question: "Can a single neuron synthesize and store more than one neurotransmitter at a time?" Evidence is accumulating that developing neuroblasts may exhibit such a property. It remains to be established if the mature mammalian neuron can store and synthesize more than one neurotransmitter at a time. Many variant clones, exhibiting a particular defect in the regulation of neuronal function, are available, and many more are being developed. Such clones may provide a useful biological tool in the study of the genetic basis for the expression of neuronal functions.

## REFERENCES

1. Davidson, R. L., Control of expression of differentiated functions in somatic cell hybrids, in: *Somatic Cell Hybridization* (R. L. Davidson and F. De La Cruz, eds.), Raven Press, New York, 1974.
2. McMorris, F. A., Kolber, A. R., Moore, B. W., and Perumal, A. S., Expression of the neuron-specific protein, 14-3-2, and steroid sulfatase in neuroblastoma cell hybrids, *J. Cell. Physiol.* **84:**473–480, 1974.
3. McMorris, F. A., and Ruddle, F. H., Expression of neuronal phenotypes in neuroblastoma cell hybrids, *Dev. Biol.* **39:**226–246, 1974.
4. Minna, J., Glazer, D., and Nirenberg, M., Genetic dissection of neural properties using somatic cell hybrids, *Nature (London) New Biol.* **235:**225–231, 1972.

5. Peacock, J. H., McMorris, F. A., and Nelson, P. G., Electrical excitability and chemosensitivity of mouse neuroblastoma × mouse or human fibroblast hybrids, *Exp. Cell Res.* **79:**199–212, 1973.
6. Yogeeswaran, G., Murray, R. K., Pearson, M. L., Sanwal, B. D., McMorris, F. A., and Ruddle, F. H., Glycosphingolipids of clonal lines of mouse neuroblastoma and neuroblastoma × L cell hybrids, *J. Biol. Chem.* **248:**1231–1239, 1973.
7. Minna, J., Nelson, P., Peacock, J., Glazer, D., and Nirenberg, M., Genes for neuronal properties expressed in neuroblastoma × L-cell hybrids, *Proc. Natl. Acad. Sci. U.S.A.* **68:**234–239, 1971.
8. Hamprecht, B., Traber, J., and Lamprecht, F., Dopamine $\beta$-hydroxylase activity in cholinergic neuroblastoma × glioma hybrid cells; increase of activity by $N^6O^{2'}$-dibutyryl adenosine 3':5'-cyclic monophosphate, *FEBS Lett.* **42:**221–226, 1974.
9. Prasad, K. N., and Hsie, A. W., Morphological differentiation of mouse neuroblastoma cells induced in vitro by dibutyryl adenosine 3':5'-cyclic monophosphate, *Nature (London) New Biol.* **233:**141–142, 1971.
10. Prasad, K. N., Differentiation of neuroblastoma cells in vitro, *Biol. Rev.* **50:**129–165, 1975.
11. Amano, T., Hamprecht, B., and Kemper, W., High activity of choline acetyltransferase induced in neuroblastoma × glia hybrid cells, *Exp. Cell Res.* **85:**399–408, 1974.
12. Greene, L. A., Shain, W., Chalazonitis, A., Breakfield, X., Minna, J., Coon, H. G., and Nirenberg, M., Neuronal properties of hybrid neuroblastoma × sympathetic ganglion cells, *Proc. Natl. Acad. Sci. U.S.A.* **72:**4923–4927, 1975.
13. Daniels, M. P., and Hamprecht, B., The ultrastructure of neuroblastoma glioma somatic cell hybrids. Expression neuronal characteristics stimulated by dibutyryl adenosine 3':5'-cyclic monophophate, *J. Cell. Biol.* **63:**691–699, 1974.
14. Chalazonitis, A., Greene, L. A., and Shain, W., Excitability and chemosensitivity properties of a somatic cell hybrid between mouse neuroblastoma and sympathetic ganglion cells, *Exp. Cell Res.* **96:**225–238, 1975.
15. Klee, W. A., Sharma, S. K., and Nirenberg, M., Opiate receptors as regulators of adenylate cyclase, *Life Sci.* **16:**1869–1874, 1975.
16. Klee, W. A., and Nirenberg, M., A neuroblastoma × glioma hybrid cell line with morphine receptors, *Proc. Natl. Acad. Sci. U.S.A.* **71:**3474–3477, 1974.
17. Sharma, S. K., Nirenberg, M., and Klee, W. A., Morphine receptors as regulators of adenylate cyclase activity, *Proc. Natl. Acad. Sci. U.S.A.* **72:**590–594, 1975.
18. Blosser, J. C., Abbott, J. R., and Shain, W., Sympathetic ganglion × neuroblastoma somatic cell hybrids with opiate receptor activity, *Fed. Proc.* **34:**713a, 1975.
19. Simon, E. J., Hiller, J. M., and Edelman, I., Stereospecific binding of the potent narcotic analgesic [³H]etorphine to rat-brain homogenate, *Proc. Natl. Acad. Sci. U.S.A.* **70:**1947–1949, 1973.
20. Pert, C. B., and Snyder, S. H., Opiate receptor binding of agonists and antagonists affected differentially by sodium. *Mol. Pharmacol.* **10:**868–879, 1974.
21. Traber, J., Fischer, K., Latzin, S., and Hamprecht, B., Morphine antagonises action of prostaglandin in neuroblastoma and neuroblastoma × glioma hybrid cells, *Nature (London)* **253:**120–122, 1975.
22. Collier, H. O. J., and Roy, A. C., Morphine-like drugs inhibit the stimulation by E prostaglandins of cyclic AMP formation by rat brain homogenate, *Nature (London)* **248:**24–27, 1974.
23. Collier, H. O. J., and Roy, A. C., Hypothesis inhibition of E prostaglandin-sensitive adenyl cyclase as the mechanism of morphine analgesia, *Prostaglandins* **7:**361–376, 1974.

24. Sharma, S. K., Klee, W. A., and Nirenberg, M., Dual regulation of adenylate cyclase accounts for narcotic dependence and tolerance, *Proc. Natl. Acad. Sci. U.S.A.* **72**:3092–3096, 1975.
25. Smith, A. A., Karmin, M., and Gavitt, J., Tolerance to the lenticular effects of opiates, *J. Pharmacol. Exp. Ther.* **156**:85–91, 1967.
26. Manner, G., Foldes, F. F., Kuleba, M., and Deery, A. M., Morphine tolerance in a human neuroblastoma line: Changed in choline acetylase and cholinesterase activities, *Experientia* **30**:137–138, 1974.
27. Birks, R., and MacIntosh, F. C., Acetylcholine metabolism of a sympathetic ganglion, *Can. J. Biochem. and Physiol.* **39**:787–827, 1961.
28. Browning, E. T., and Schulman, M. P., [$^{14}$C]Acetylcholine synthesis by cortex slices of rat brain, *J. Neurochem.* **15**:1391–1405, 1968.
29. Ansell, G. B., and Spanner, S., The long-term metabolism of the ethanolamine moiety of rat brain myelin phospholipids, *J. Neurochem.* **15**:1371–1373, 1968.
30. Yamamura, H. I., and Snyder, S. H., Choline: High-affinity uptake by rat brain synaptosomes, *Science* **178**:626–628, 1972.
31. Dowdall, M. J., and Simon, E. J., Comparative studies on synaptosomes: Uptake of (N-Mc-34) choline by synaptosomes from squid optic lobes, *J. Neurochem.* **21**:969–982, 1973.
32. Massarelli, R., Ciesielski-Treska, J., Ebel, A., and Mandel, P., Choline uptake in neuroblastoma cell cultures: Influence of ionic environment, *Pharmacol. Res. Commun.* **5**:397–406, 1973.
33. Massarelli, R., Ciesielski-Treska, J., Ebel, A., and Mandel, P., Kinetics of choline uptake in neuroblastoma clones, *Biochem. Pharmacol.* **23**:2857–2865, 1974.
34. Massarelli, R., Sensenbrenner, M., Ebel, A., and Mandel, P., Incorporation de choline dans une culture de cellules du recherche d'embryou de poulet, *J. Physiol. (France)* **67**:292A, 1973.
35. Stefanović, V., Massarelli, R., Mandel, P., and Rosenberg, A., Effect of cellular desialylation on choline high affinity uptake and ecto-acetylcholinesterase activity of cholinergic neuroblasts, *Biochem. Pharmacol.* **24**:1923–1928, 1975.
36. Lanks, K., Somers, L., Papirmeister, B., and Yamamura, H., Choline transport by neuroblastoma cells in tissue culture, *Nature (London)* **252**:476–478, 1974.
37. Richelson, E., and Thompson, E. J., Transport of neurotransmitter precursors into cultured cells, *Nature (London) New Biol.* **241**:201–204, 1973.
38. Zwiller, J., Ciesielski-Treska, J., Mack, G., and Mandel, P., Uptake of noradrenaline by an adrenergic clone of neuroblastoma cells, *Nature (London)* **254**:443–444, 1975.
39. Iversen, L. L., Uptake mechanisms for neurotransmitter amines, *Biochem. Pharmacol.* **23**:1927–1935, 1974.
40. Kimelberg, H. K., Active potassium transport and [$Na^+ + K^+$]ATPase activity in cultured clioma and neuroblastoma cells, *J. Neurochem.* **22**:971–976, 1974.
41. Rotman, A., Daly, J. W., Creveling, C., and Breakefield, X. O., Uptake and binding of dopamine and 6-hydroxydopamine in murine neuroblastoma and fibroblast cells, *Biochem. Pharmacol.* **25**:383–388, 1976.
42. Angeletti, P. U., and Levi-Montalcini, R., Cytolytic effect of 6-hydroxydopamine on neuroblastoma cells, *Cancer Res.* **30**:2863–2869, 1970.
43. Prasad, K. N., Effect of dopamine and 6-hydroxydopamine on mouse neuroblastoma cells *in vitro*, *Cancer Res.* **31**:1457–1460, 1971.
44. Axelrod, J., The metabolism, storage, and release of catecholamines, *Recent Prog. Horm. Res.* **21**:597–622, 1965.
45. Curtis, D. R., and Johnston, G. A. R., Amino acid transmitter, in: *Handbook of*

   *Neurochemistry*, Vol. 4 (A. Lajtha, ed.), pp. 115–134, Plenum Press, New York, 1970.

46. Henn, F. A., and Hamberger, A., Glial cell function: Uptake of transmitter substances, *Proc. Natl. Acad. Sci. U.S.A.* **68**:2686–2690, 1971.

47. Hutchison, H. T., Werrbach, K., Vance, C., and Haber, B., Uptake of neurotransmitters by clonal lines of astrocytoma and neuroblastoma in culture. I. Transport of γ-aminobutyric acid, *Brain Res.* **66**:265–274, 1974.

48. Schubert, D., The uptake of GABA by clonal nerve and glia, *Brain Res.* **84**:87–98, 1975.

49. DeLellis, R. A., Rabson, A. S., and Albert, D., The cytochemical distribution of catecholamines in the C-1300 murine neuroblastoma, *J. Histochem. and Cytochem.* **18**:913–914, 1970.

50. Schubert, D., Humphreys, S., Baroni, C., and Cohn, M., *In vitro* differentiation of a mouse neuroblastoma, *Proc. Natl. Acad. Sci. U.S.A.* **64**:316–323, 1969.

51. Narotzky, R.N and Bondareff, W., Biogenic amines in cultured neuroblastoma and astrocytoma cells, *J. Cell Biol.* **63**:64–70, 1974.

52. Prasad, K. N., Mandal, B., Waymire, J. C., Lees, G. J., Vernadakis, A., and Weiner, N., Basal levels of neurotransmitter synthesizing enzymes and effect of cyclic AMP agents on the morphological differentiation of isolated neuroblastoma clones, *Nature (London) New Biol.* **241**:117–119, 1973.

53. Schubert, D., Heinemann, S., Carlisle, W., Tarikas, H., Kimes, B., Patrick, J., Steinbach, J. H., Culp, W., and Brandt, B. L., Clonal cell lines from the rat central nervous system, *Nature (London)* **249**:224–227, 1974.

54. Cicero, T. J., Cowan, W. M., Moore, B. W., and Suntzeff, V., The cellular localization of the two brain specific proteins S-100 and 14-3-2, *Brain Res.* **18**:25–34, 1970.

55. Perez, V. J., Olney, J. W., Cicero, T. J., Moore, B. W., and Bahn, B. A., Wallerian degeneration in rabbit optic nerve: Cellular localization in the central nervous system of S-100 and 14-3-2 proteins, *J. Neurochem.* **17**:511–519, 1970.

56. Packman, P. M., Blomstrand, C., and Hamberger, A., Disc electrophoretic separation of proteins in neuronal, glial and subcellular fractions from cerebral cortex, *J. Neurochem.* **18**:479–487, 1971.

57. Haglid, K., Carlsson, C. A., and Stavrou, D., An immunological study of human brain tumors concerning the brain specific proteins S-100 and 14-3-2, *Acta Neuropathol. (Berlin)* **24**:187–196, 1973.

58. Hydén, H., and McEwen, B., A glial protein specific for the nervous system, *Proc. Natl. Acad. Sci. U.S.A.* **55**:354–358, 1966.

59. Sviridov, S. M., Korochkin, L. I., Ivanov, V. N., Maletskaya, E. I., and Bakhtina, T. K., Immunohistochemical studies of S-100 protein during postnatal ontogenesis of the brain of two strain of rats, *J. Neurochem.* **19**:713–718, 1972.

60. Sensenbrenner, M., Differentiation of cells in dissociated cell culture, in: *Cell, Tissue and Organ Cultures in Neurobiology* (S. Fedoroff and L. Hertz, eds.), pp. 191–213, Academic Press, New York, 1977.

61. Walum, E., and Eström, A., Kinetics of 2-deoxy-D-glucose transport into cultured mouse neuroblastoma cells, *Exp. Cell. Res.* **97**:1–8, 1976.

62. Prasad, K. N., Sahu, S. K., and Kumar, S., Relationship between cyclic AMP level and differentiation of neuroblastoma cells in culture, in: *Differentiation and Control of Malignancy of Tumor Cells* (W. Nakahara, T. Ono, T. Sugimura, and H. Sugano, eds.), pp. 287–309, University of Tokyo Press, Tokyo, 1974.

63. Edström, A., Kanje, M., and Walum, E., Density dependent inhibiton of 2-deoxy-D-glucose uptake into glioma and neuroblastoma cells in culture, *Exp. Cell Res.* **97**:8–15, 1976.

64. Newburgh, R. W., and Rosenberg, R. N., Effect of norepinephrine on glucose metabolism in glioblastoma and neuroblastoma cells in cell culture, *Proc. Natl. Acad. Sci. U.S.A.* **69**:1677–1680, 1972.
65. Passonneau, J. V., and Crites, S. K., Regulation of glycogen metabolism in astrocytoma and neuroblastoma cells in culture, *J. Biol. Chem.* **251**:2015–2022, 1976.
66. Guth, L., and Watson, P. K., A correlated histochemical and quantitative study on cerebral glycogen after brain injury in the rat, *Exp. Neurol.* **22**:590–602, 1968.
67. Passonneau, J. V., and Lowry, O. H., in: *Recent Advances in Quantitative Histo- and Cytochemistry* (U. C. Dubach and U. Schmidt, eds.), pp. 198–212, Hans Huber, Bern, 1971.
68. Wintzerith, M., Ciesielski-Treska, J., Dierich, A., and Mandel, P., Comparative investigation of free nucleotides in two neuroblastoma clonal cell lines, *J. Neurochem.* **26**:205–207, 1976.
69. Mandel, P., and Edel-Harth, S., Free nucleotides in the rat brain during post-natal development, *J. Neurochem.* **13**:591–595, 1966.
70. Schubert, D., Carlisle, W., and Look, C., Putative neurotransmitters in clonal cell lines, *Nature (London)* **254**:341–343, 1975.
71. Saifer, A., Comparative study of various extraction methods for the quantitative determination of free amino acids from brain tissue, *Analyt. Biochem.* **40**:412–423, 1971.
72. Hokin, L. E., Dynamic aspects of phospholipids during protein secretion, *Int. Rev. Cytol.* **23**:187–208, 1968.
73. Lapetina, E. G., and Michell, R. H., Phosphatidylinositol metabolism in cells receiving extracellular stimulation, *FEBS Lett.* **31**:1–10, 1973.
74. Honkin, L. E., Phospholipid metabolism and functional activity of nerve cells, in: *Structure and Function of Nervous Tissue*, Vol. 3 (G. H. Bourne, ed.), pp. 161–184, Academic Press, New York, 1969.
75. Schacht, J., and Agranoff, B. W., Effects of acetylcholine on labeling of phosphatidate and phosphoinositides by [$^{32}$P]orthophosphate in nerve ending fractions of guinea pig cortex, *J. Biol. Chem.* **247**:771–777, 1972.
76. Yagihara, Y., Bleasdale, J. E., and Hawthorne, J. N., Effects of acetylcholine on the incorporation of [$^{32}$P]orthophosphate *in vitro* into the phospholipids of subsynaptosomal membranes from guinea-pig brain, *J. Neurochem.* **21**:173–190, 1973.
77. Hokin, M. R., Effect of norepinephrine of $^{32}$P incorporation into individual phosphatides in slices from different areas of the guinea pig brain, *J. Neurochem.* **16**:127–134, 1969.
78. Abdel-Latif, A. A., Yau, S. J., and Smith, J. P., Effect of norepinephrine and other agents on $^{32}$Pi incorporation into phospholipids and phosphoproteins of rats and guinea pig brain slices, *Trans. Am. Soc. Neurochem.* **5**:66a, 1974.
79. Berg, G. R., and Klein, D. C., Norepinephrine increases the [$^{32}$P]labelling of a specific phospholipid fraction of post-synaptic pineal membranes, *J. Neurochem.* **19**:2519–2532, 1972.
80. Eichberg, J., Shein, H. M., Schwartz, M., and Hauser, G., Stimulation of $^{32}$Pi incorporation into phosphatidylinositol and phosphatidylglycerol by catecholamines and β-adrenergic receptor blocking agents in rat pineal organ cultures, *J. Biol. Chem.* **248**:3615–3622, 1973.
81. Eichberg, J., Shein, H. M., and Hauser, G., Phospholipid metabolism in cultured neuroblastoma and glioma cells incubated with carbamylcholine and norepinephrine, *J. Neurochem.* **24**:67–70, 1975.
82. Martin, S. E., Mouse brain antigen detected by rat anti-C-1300 antiserum, *Nature (London)* **249**:71–73, 1974.

83. Terman, D. S., Stewart, I., Tavel, A., and Kirch, D., Localization of neuroblastoma *in vivo* with tumor specific antibodies, *Cancer Res.* **35:**1761–1766, 1975.
84. Bertolini, L., Diamond, L., and Revoltella, R., Modification of growth of neuroblastoma cells in syngenic mice with aldehyde-treated neuroblastoma cells, *Cancer Res.* **36:**2111–2112, 1976.
85. Byfield, J. E., Zerubavel, R., and Fonkalsrud, E. W., Murine neuroblastoma cured *in vivo* by an antibody-dependent cellular cytotoxicity reaction, *Nature (London)* **264:**783–785, 1976.
86. Akeson, R., and Herschman, H. R., Modulation of cell surface antigens of a murine neuroblastoma, *Proc. Natl. Acad. Sci. U.S.A.* **71:**187–191, 1974.
87. Jorgensen, A. O., Subrahmanyan, L., Turnbull, C., and Kalnins, V. I., Localization of the neurofilament protein in neuroblastoma cells by immunofluorescent staining, *Proc. Natl. Acad. Sci. U.S.A.* **73:**3192–3196, 1976.
88. Carlin, S. C., Rosenberg, R. N., VandeVenter, L., and Friedkin, M., Quinazoline antifolates as inhibitors of growth, dihydrofolate reductase, and thymidylate synthetase of mouse neuroblastoma cells in culture, *Mol. Pharmacol.* **10:**194–203, 1974.
89. Sensenbrenner, M., Lodin, Z., Treska, J., Jacob, M., Kage, M. P., and Mandel, P., The cultivation of isolated neurons from spinal ganglia of chick embryo. *Z. Zellforsch. Mikrosk. Anat.* **98:**538–549, 1969.
90. Lodin, Z., Booher, J., and Kasten, F. H., Phase-contrast cinematographic study of dissociated neurons from embryonic chamber chick dorsal root ganglia cultured in the rose chamber, *Exp. Cell Res.* **60:**27–39, 1970.
91. Varon, S., and Raiborn, C., Dissociation, fractionation, and culture of chick embryo sympathetic ganglionic cells, *J. Neurocytol.* **1:**211–221, 1972.
92. Luduena, M. A., Nerve cell differentiation *in vitro*, *Dev. Biol.* **33:**268–284, 1973.
93. Sensenbrenner, M., and Mandel, P., Behaviour of neuroblasts in the presence of glial cells, fibroblasts and meningeal cells in culture, *Exp. Cell Res.* **87:**159–167, 1974.
94. Patterson, P. H., and Chun, L. L. Y., The influence of non-neuronal cells on catecholamine and acetylcholine synthesis and accumulation in cultures of dissociated sympathetic neurons, *Proc. Natl. Acad. Sci. U.S.A.* **71:**3607–3610, 1974.
95. Stefanovic, V., Ciesielski-Treska, J., Ebel, A., and Mandel, P., Neuroblasts–glia interaction. The effect of co-cultivation upon ecto-ATPase activity of neuroblastoma and glioma cell, *Exp. Cell Res.* **98:**191–203, 1976.
96. Giller, E. L., Schrier, B. K., Shainberg, A., Fisk, H. R., and Nelson, P. G., Choline acetyltransferase activity is increased in combined cultures of spinal cord and muscle cell from mice, *Science* **182:**588–589, 1973.

# SOME ASPECTS OF DEVELOPMENT OF NERVOUS TISSUE

## MORPHOGENESIS

Vertebrate spinal and autonomic ganglia are derived from embryonic neural crest cells.[1] This unique population of cells is also the developmental antecedent of a large number of other differentiated cell types, including neuroendocrine tissues such as the adrenal medulla and the calcitonin-producing ("C") cells,[2] and the glial (Schwann sheath and satellite) cells of the peripheral nervous system. Integumental pigment cells and cells that form skeletal and connective tissue of the head and face[3] are also derivatives of the embryonic neural crest.

There is now considerable evidence that crest cells are pluripotent in early migratory stages and, therefore, that the initial migration and subsequent differentiation of these cells is at least partially regulated by environmental conditions. The most compelling support for this conclusion is provided by the results of heterotopic grafting operations in which crest from one embryonic region is transplanted to another location. For example, it has been demonstrated that the precise migratory pathways followed by crest cells in the embryo are characteristic of particular axial levels and that, when labeled crest from one axial level of an embryo is grafted heterotopically into an unlabeled host, the pattern of cell migration corresponds to that characteristic of the graft site rather than that of the origin of the grafted crest cells.[4] To some extent, the migratory pathways of crest cells may be specified by cues in their immediate environment.[1,3,5]

The differentiation of neural crest cells into particular neuronal derivatives is also influenced by local environmental conditions. This was shown[6,7] by means of heterotopic, interspecific (chick–quail) grafting procedures that utilized differences in nuclear staining properties to distinguish cells of chick or quail origin.[8] Crest of the posterior metencephalic (postotic) axial region of avian embryos normally contributes some of its cells to the adrenergic neurons of the sympathetic chain and the adrenal medulla. When neural tubes from these two axial levels were exchanged in a series of grafting operations, the crest cells from the donors not only migrated but also differentiated as would crest cells normally arising in the host area. Thus, the crest cells differentiate in response to environmental cues encountered during migration rather than in response to earlier cues.

Crest cell differentiation can also be influenced by environmental factor *in vitro*. For example, regardless of their normal developmental capabilities or fates, when avian or amphibian crest cells are cultured without specific interactions with other embryonic components,[9-11] most of the cultured crest cells will make pigment. However, if cranial neural crest is cultured in association with either pharyngeal endoderm[12] or extracellular products produced *in vitro* by retinal pigment epithelium,[13] the cultures produce cartilage, a tissue formed *in vivo* by head crest. Moreover, sympathetic neurons will differentiate if trunk crest cells from chicken embryos are permitted to associate closely *in vitro* with somitic mesenchyme[14] that has been "conditioned" by suitable interactions with ventral neural tube tissue.[15]

Differentiation of spinal ganglion neurons from cultured crest cells has not yet been demonstrated, although such neurons can be maintained in culture once the initial neurogenic events within spinal ganglia have occurred.[16] However, metaplasia in cultures of spinal or sympathetic ganglia provides additional evidence that crest cells may be pluripotent and that local environmental factors continue to play an important role in eliciting the expression of particular crest phenotypes. It has been reported that pigment cells frequently appear in cultures of embryonic spinal ganglia.[17] It has been suggested that the degree of dispersion (or conversely, the extent of association) of crest-derived cells plays a role in determining the differentiative pathway they follow.[18,19] It has been shown[20] that catecholamine-containing sympathetic neurons can differentiate under certain conditions from cultured spinal ganglia, although they would never normally do so *in vivo*.

It has also been reported[21] that primary sympathetic neurons from neonatal rat superior cervical ganglia grown in the absence of other (nonneuronal) cell types normally found in the ganglion can synthesize and store the transmitter noradrenalin. In mixed cultures of these neurons and nonneuronal cells, however, the level of acetylcholine increases markedly, suggesting that metaplasia of crest-derived cells depends on precise environmental interactions. The role of nonneuronal cells in the differentiation of ganglia is clearly complex. They have been reported to enhance neuron survival when dissociated spinal ganglion neurons are cultured in the absence of nerve growth factor (NGF).[22]

The onset and extent of crest cell migration appear to be regulated by changes in the embryonic tissue environment.[23] Thus, it has been shown[24] that cranial crest cell migration is preceded by the appearance in the embryo of a cell-free space filled with an extracellular matrix rich in hyaluronic acid. Likewise, trunk neural crest cells begin to migrate from their origin at the top of the neural tube in close association with components of extracellular matrix. Initially, migrating crest cells associate closely with basement membranes of the neural tube and the overlying ectoderm.[25] Subsequently, crest cells enter and are surrounded by a matrix of amorphous extracellular material. The glycosaminoglycan (GAG) components of this extracellular material (ECM) may be characterized qualitatively using both their specific interactions with the cationic dye Alcian blue at various magnesium ion concentrations,[26] and its susceptibility to enzymes that selectively degrade GAG components of the matrix. The relative amounts of Alcian blue bound in histological sections may be estimated by microspectrophotometry, using the two-wavelength method.[27]

## Initiation of Migration

Transverse sections of avian and murine embryos were examined at the stage and axial level where trunk crest cell migration had just begun. In the region of migrating crest cells there is considerable material that binds Alcian blue under conditions that allow all GAG to stain (0.1 M $MgCl_2$; pH 2.6). Microspectophotometry[27] reveals that approximately 60% of this staining is lost when histological sections are pretreated with *Streptomyces* hyaluronidase that specifically degrades hyaluronic acid.[28] In contrast, the amount of Alcian blue stain in the somitic region of these same embryo sections appears to be about one-third the amount found in the region of the neural crest, and only about 30% of this stain is susceptible to hyaluronidase treatment.

Likewise, when such sections are stained under conditions that are thought to limit reaction of the Alcian blue to sulfated material (0.3 M MgCl$_2$; pH 5.8),[26] microspectrophotometry indicates about twice as much stain in the crest region as in the somite. The proportion of stain sensitive to chondroitinase ABC, which degrades chondroitin sulfate and dermatan sulfate, is about the same in the two regions.

From these results the following conclusions were made[23]: (1) There is more GAG in the region where crest cells begin to migrate than in nearby somite; (2) the proportion of hyaluronidase-sensitive stain (hyaluronic acid) in the region of the neural crest is more than twice that in the somite; and (3) the proportion of chondroitinase ABC-sensitive stain (chondroitin sulfate and dermatan sulfate) is about the same in the two regions.

## Restriction of Migration

Heterochronic grafts, in which crest cells were placed into developmentally older host regions, suggested that the extent of crest cell migration in the region of the somite was progressively restricted, and that this restriction was related to a change in the ability of the somitic region to support crest cell migration.[29] It was of considerable interest, therefore, that when transverse sections of older embryos were examined, a different Alcian blue staining pattern was observed, which showed striking correlations with changes in crest cell morphogenetic behavior. Thus, photometric comparisons of Alcian blue staining in the region dorsolateral to the neural tube, and the region between the ectodermal epithelium and the dermatome, demonstrated that the extracellular material included both hyaluronic acid and sulfated GAG. Likewise, in somitic mesenchyme, Alcian blue-staining matrix components were present, although in reduced amounts compared to the above regions. However, where crest cells had begun to coalesce to form spinal ganglia, significantly less Alcian blue staining was observed compared to the other regions.

The pattern that emerges seems clear: early crest cell migration proceeds into extracellular spaces containing a variety of materials, including collagen, proteoglycans, and hyaluronic acid. Where crest cells remain dispersed, as appears to be the case of prospective melanoblasts under the ectodermal epithelium, the crest cells are surrounded by Alcian blue staining material. Where crest cells coalesce to form spinal ganglia, however, this close cell association seems to be correlated with a relative decrease of GAG in the extracellular matrix.[23] It is not yet known whether the change in extracellular material actually precedes coalescence of crest cells into ganglia, but it is of some interest that the apparent loss in ability of the somitic region to support crest cell migration revealed by heterochronic grafting of crest[23-29] corresponds roughly in stage of development and in embryonic region (stage 25; midthoracic level of the chicken embryo) to the decrease in the proportion of hyaluronic acid in axial mesenchyme.[30]

## Synthesis of GAG

The sources of GAG in the extracellular spaces are not known; nor is it known what mediates the changes in the amounts and proportions of GAG in different axial regions and at different times. However, several lines of evidence suggest that crest cells themselves make at least some of the hyaluronate found in the matrix. Thus, it has been found that primary cultures of crest cells from explanted quail neural tubes cultured by standard methods[10,11] incorporated [³H]glucosamine largely into hyaluronic acid. A small amount of sulfated GAG may also be made by crest cells *in vitro*. Similar results, using slightly different methods, were obtained[23] for cultured cranial neural crest cells. In contrast, cultures of somitic mesenchyme from the same embryos incorporate a larger percentage (up to 60%) of labeled precursors into chondroitin sulfate and other sulfated GAG, but seem to make hyaluronic acid as well. The pattern of synthesis by these tissues correlates well with observed Alcian blue staining that indicated the composition and relative proportions of GAG in the regions of crest cell migration and the somite.[23]

## Cell Association and Gangliogenesis

Coalescence of crest cells to form spinal ganglia within somitic mesenchyme is clearly accompanied by a local reduction of Alcian blue staining material. Loss of this extracellular material very likely favors cell association, which in turn appears to be involved in neuronal differentiation of crest cells.[19,22]

When it is excised from the embryo, the nascent sensory ganglion already contains several different cell types. These include large and small neurons and neuroblasts, as well as supportive (satellite and Schwann sheath) cells and their precursors. In addition, the ganglion is utimately surrounded by a capsule of meningeal fibroblasts that probably arise from somitic mesenchyme.[23] In light of the cellular metaplasia exhibited by cultured ganglia,[17,18,23] it is of considerable interest to know the source(s) of the adventitious cell types and the environmental conditions that promote their appearance.

## Source of Pigment Cells in Ganglion Cultures

Cultures of 5-day-old spinal ganglia contain a characteristic population of small stellate cells with intensely staining nuclei that strikingly resembles cultured crest cells.[11] These cells are clearly distinguishable from larger, fibroblastic (meningeal) cells and from neuronal cells with which they are closely associated. Moreover, they are morphologically similar to the melanosome-containing pigment cells that appear in prolonged cultures of these 5-day-old ganglia. It seems possible, therefore, that these small stellate cells are in fact the antecedents of the adventitious pigment cells that appear in cultured spinal ganglia. Since similar small cells with dark-staining nuclei are observed to associate closely with nerve fiber bundles *in vivo* and *in vitro*, it was possible to test for their ability to make pigment by culturing peripheral nerve removed from stage 26 (5-day-old) chick embryos. Such explants contain the small stellate cells associated with nerve fibers, and some fibroblastic cells from the fiber sheath, but are usually devoid of nerve cell bodies or neuroblasts. In these cultures, the fibers soon degenerate, leaving behind on the substratum a population of the small stellate cells that resemble the cells observed in cultures of spinal ganglia. After several days in a suitable culture medium, some of these cells undergo melanogenesis.[23] These results suggest that at least one source of the pigment cells that appear in cultured spinal ganglia is the population of small stellate cells that normally associate with nerve cell bodies and fibers and that appear to be supportive (satellite or Schwann sheath) cells or their precursors.

## Regulation of Melanogenesis in Cultured Ganglia

Close examination of ganglion cultures undergoing melanogenesis reveals that pigment cells differentiate only in populations of small stellate cells that are not associated with neurons or fibers. To investigate further the relationship between cell association and melanogenesis, several types of experiments were performed. First, a variety of culture conditions were utilized that favored the dissociation of supportive cells from neurons. These conditions included (1) cultures on plastic substrata of enzymatically dissociated ganglia in a "permissive" medium containing serum and embryo extract, and (2) cultures of intact young ganglia on plastic or collagen substrata in permissive media. Under these conditions, numerous small stellate cells were observed that were not associated with neurons or their fibers, and many of these cells formed pigment.

In contrast, when ganglia were cultured on fibroblastic (meningeal) substrata or on agar substrata, both of which favored the maintenance of association between nerve cells of fibers and supportive cells, melanogenesis did not occur, even though the cultures were maintained in a permissive medium. Results from cultures of ganglia partially on a substratum of meningeal fibroblasts and partially on plastic were particularly instructive. Under nutritional conditions that are probably identical for all cells (although very short-range cellular effects may alter nutritional microenvironments), pigment appeared only in regions of the ganglion situated on plastic. On plastic substrata, many of the small stellate (supportive) cells seem to have deserted the neurons and fibers in favor of the plastic, whereas most of the supportive cells remained associated with nerves when the alternative substratum was a meningeal cell monolayer.

Finally, when ganglia were cultured in the initial absence of NGF activity and then transferred to pigment-permissive medium, melanogenesis among the surviving cells, which had many fewer neurons left with which to associate, was enhanced.[19]

These results are all compatible with the idea that some of the supportive cells had undergone metaplasia and that association between supportive cells and neurons was inimical to this process. They further suggest that when supportive cells, which normally associate closely with nerve cell bodies or fibers, are given access to alternative substrate *in vitro*, the choice is not made at random. The preference of supportive cells or their precursors for neurons over meningeal fibroblasts suggests that meninges may normally promote gangliogenesis *in vivo* by preventing association of supporting cells with each other, thereby preventing such tissue disruption as occurs on artificial substrata *in vitro*.

These results support the idea that crest cells are pluripotent initially and that some remain so for a considerable period during development. Further, the results suggest how one aspect of cell social behavior—cell association—might affect cell differentiation and how this, in turn, might be regulated by environmental factors: crest cells migrate into embryonic tissue spaces filled with extracellular material composed of collagens, proteoglycans, and glycosaminoglycans. Hyaluronic acid, an important GAG constituent that is plentiful in regions where early crest migration occurs,[24] may be produced, in part, by crest cells themselves immediately preceding the onset of migration. The hyaluronate-rich extracellular matrix provides a medium that seems to favor crest cell proliferation and locomotion and tends to keep the cells apart. Later, there is a change in the ability of the somitic regions to support crest cell migration that may be correlated with a decrease in the proportion of hyaluronic acid in the extracellular matrix. In any case, the coalescence of crest cells into coherent ganglia is seen to be accompanied by a dramatic local reduction in Alcian blue staining material. The cell associations that result from this coalescence favor neuron and supportive cell differentiation and survival. Conversely, the disruption of these associations that sometimes occurs when attractive alternative substrata are presented *in vitro*, leads still-pluripotent crest-derived cells to embark on other differentiative pathways.[24]

## NEURONAL INTERACTION

Although it has been established that development of innervation is necessary for normal maturation of organs such as skeletal muscle, little is known about the mechanisms by which developing neurons interact to regulate the maturation of one another.[31] The superior cervical ganglion (SCG) in the neonatal mouse and rat has been employed as a model system with which to study the regulation of ontogeny of presynaptic cholinergic nerves and postsynaptic adrenergic neurons.[31] During postnatal development, presynaptic CAT activity increases 30- to 40-fold, whereas postsynaptic TH activity rises 6- to 8-fold. Transection of the presynaptic cholinergic nerves innervating the SCG prevents the normal development of TH activity and the normal accumulation of TH enzyme molecules in each postsynaptic neuron. The transsynaptic regulation of TH development is apparently mediated by acetylcholine and postsynaptic depolarization, since pharmacologic ganglionic blockade also prevents normal maturation. Ganglion decentralization also prevents the normal maturation of adrenergic nerve terminals, and the development of end-organ innervation by SCG. Consequently, transsynaptic factors regulate the ontogeny of adrenergic terminals as well as perikarya.

Moreover, normal efferent as well as afferent connections are apparently required for sympathetic development, since removal of salivary glands and orbital contents, target organs of the SCG, in neonates also prevents TH development in the ganglia.

The postsynaptic neuron contributes to the development of presynaptic cholinergic fibers in SCG. Selective destruction of adrenergic neurons in neonatal mice with either 6-OHDA or antiserum to nerve growth factor pervents the normal maturation of CAT activity in presynaptic terminals of SCG. Thus, presynaptic and postsynaptic cells appear to exert reciprocal regulatory influences during ontogeny.


## CELL RECOGNITION

One of the most striking features of the vertebrate central nervous system is the precise spatial arrangement of neuronal cell bodies and processes. Neuronal somata are not scattered randomly but instead are grouped into laminae or nuclei that are highly characteristic of a particular region of the brain. The spatial arrangement of axons is also highly ordered. Axons run in defined tracts and terminate on restricted populations of target neurons. Populations of axons have the added property of forming ordered two-dimensional arrays, of which the retinotopic projection of the optic nerve is a well-studied example. The molecular basis of these kinds of pattern formation in the nervous system is not known, but it is reasonable to speculate that surface cell–cell receptors might play an important role; consequently, several authors have proposed mechanisms involving complementary receptors and ligands on "appropriate" pairs of plasma membranes.[32]

Roth, Roseman, and their co-workers have provided direct evidence for cell–cell receptors on the surface of embryonic cells by showing that single cells derived by dissociating embryonic tissues bind preferentially to aggregates or monolayers of homologous cells.[33,34] These demonstrations of cell–cell receptors raise two related questions: first, "What is the molecular nature of the receptors and ligands involved?" and second, "What role do they play in normal development?" Evidence suggests[35] that cell-surface receptors occur on the surface of developing brain cells and that by functional criteria there are at least seven types of receptor. An active component of cell–cell recognition can be solubilized and partially purified. Quantitative studies of cell–cell adhesion demonstrate[35] a gradient of adhesive specificity along the dorsoventral axis of the developing retina.

## CELL ADHESION

An analysis of the mechanism of adhesion among vertebrate cells is of central importance in understanding tissue formation during embryogenesis. Although the interaction between cell surfaces has been described in detail at the morphological[36,37] and ultrastructural[38] levels, the molecular mechanism of cell adhesion remains unknown. To solve this problem, a number of workers have attempted to isolate molecules or membrane fragments that can enhance reaggregation of dissociated cells or that demonstrate a tissue-specific affinity for the surface of intact cells.[39-41] In each case, active molecules or factors have been described, but the relationship of these substances to each other or to the mechanism of cell adhesion has not been established.

The data on initial binding among individual brain and retinal cells suggest that the molecular mechanism of cell adhesion is the same in retinal and brain cells.[42] Not only can cells from these two tissues bind to each other as well as to themselves, but in all cases binding can be inhibited by antibodies to the same molecule. The binding properties of each cell type, however, did vary with respect to developmental age, indicating that there is a temporal program for cell adhesion that is different for retinal and brain cells.

Immunoprecipitation of cell surface molecules by the antibodies that block cell adhesion has been used for provisional identification of the components that bind the antibodies that prevented cell–cell ligation. The antibodies that completely blocked adhesion precipitated a cell surface protein with a molecular weight of 150,000 from NP40 extracts of cells grown in suspension. On the other hand, antibodies to F1 fractions, which only weakly blocked adhesion, precipitated large (200,000–280,000 molecular weight) cell surface proteins from NP40 extracts of monolayer cells but not from extracts of suspension cells.

Because the antibodies that strongly block adhesion and precipitate the 150,000-molecular-weight component were made against fragments of F1, the precipitation of higher-molecular-weight polypeptides by anti-F1 fractions suggests the possibility that the 150,000-molecular-weight molecule may be derived from or related to a larger precursor. Furthermore, the fact that the antibodies precipitating the 150,000-molecular-weight species block adhesion completely, whereas the anti-F1 antibodies are relatively ineffective in this respect, suggest that inhibition of cell ligation occurs through an inactivation of the 150,000-molecular-weight component by antibody. These conclusions, of course, rely on the specificity of the antibodies obtained against F1 and its fragments. Inasmuch as the effect of a minor contaminant in the antigens used to produce these antibodies could be magnified in either the inhibition of adhesion or the identification of cell surface molecules, it will be necessary to do further structural and functional experiments to exclude this possibility. With these reservations in mind, a working hypothesis for the mechanism of cell adhesion that assumes that the 150,000-molecular-weight molecules on the cell surface are directly involved in the formation of cell–cell bonds, which can be inhibited by the binding of antibody specific to that molecule, is proposed. In addition, there is the possibility that this molecule is derived from a large precursor by proteolytic cleavage. The 150,000-molecular-weight polypeptide could function by being self-complementary so that a cell–cell bond would be a symmetrical dimer. Alternatively, this molecule may bind to one or more unrelated molecules. It has been found in preliminary experiments that incubation of only one of the two cells involved in each binding event with monovalent Fab' antibody is sufficient to prevent adhesion. This result is more compatible with mechanisms in which two molecules having the same antigenic determinant are involved in the formation of each cell–cell bridge.

Whatever the actual mechanism of cell–cell adhesion, it will be important to explain the temporal specificity of cell adhesion during embryogenesis in terms of molecular events. Some studies on the binding of neural cells to nylon fibers coated with the antibodies that block adhesion indicate that a temporal variation occurs in this cell-fiber interaction that parallels the variation seen in direct cell-cell binding.[43] This raises the possibility that adhesion may in part be modulated by changes in the amount, structure, or dynamic properties of the 150,000-molecular-weight cell-surface protein.

## NEURONAL DEATH

In the vertebrate nervous system the number of neurons is remarkably constant. There appears to be a precise control mechanism which regulates cell number during either neurogenesis or the period of neuronal maturation (see review, reference 44). The fact that degeneration of neuronal cells occurs during development of the nervous system has been known for a long time, but only in the past decade or so has a great deal of attention been given to this problem.[46,46] It has been found[48-50] that there is normally a considerable overproduction of neurons in the motor columns of the cord and in the spinal ganglia of amphibia, as revealed by the fact that the number of neurons present in both regions at early larvae stages is

appreciably greater than the number recognizable after metamorphosis. Serial counts of the number of cells indicate that there is a
marked loss in both the motor columns and the spinal ganglia coinci-
dent with the onset of limb movements, and during this critical
period (between stages 54 and 59 in *Xenopus*) large numbers of degenerating cells can be seen in both regions. It has been estimated[45]
that at least eight neurons must degenerate for every one that sur-
vives to maturity. This implies that during the larval period there
must be an appreciable "turnover" of cells in the amphibian nervous
system such that new neurons are continually being added to the
population while others are degenerating. There can be little doubt
that cell death is a major factor determining neuron number in the
anuran nervous system.[44] However, since there appears to be a
number of important differences in the pattern of neurogenesis between amphibians and amniotes, it is not clear whether one can
generalize from these observations of cell loss in the amphibian spinal
cord and ganglia, and the question remains whether cell death plays a
significant morphogenetic role in the nervous system of higher vertebrates. As yet, too little is known about the development of the
mammalian nervous system from this point of view. There are a few
scattered reports in the literature which suggest that at least some
degree of cell loss is a normal concomitant of neural development in
mammals. For example, it has been reported[47] that a progressive loss
of cells from the motor columns of the ventral horn of normal mice,
amounting to about 30%, occurs between the second and the twelfth
day postnatally. It has been shown[44,46] that the formation of discrete
ventral horns at the nonlimb levels of the mouse spinal cord is due to
the morphogenetic death of a substantial number of neurons. A very
substantial body of evidence suggests that, during the development
of the avian nervous system, cell death occurs rather widely and plays
a critical morphogenetic role in shaping many neural centers.[44,48]
However, the following questions remain to be answered:

1. Is this a general phenomenon in the sense that it is common to
   all nervous systems, vertebrate and invertebrate, and affects
   all, or most, neural populations?
2. Even if it is not a universal phenomenon, why should there be
   such a gross overproduction of nerve cells in so many neural
   centers, and is the cell death that occurs truly programmed or
   is it largely fortuitous?
3. Are most peripheral effects on early neural development due
   to the regulation of neuronal death, and if so, to what extent

can the size of a neuronal population be increased by reducing the level of cell loss?
4. What is the signal that triggers the death of so many cells, and how do other cells in the same population escape this influence?

## Significance of Trophic Influences

Trophic influences may be defined as those factors concerned with the growth, maintenance, and regression of nerve cells either as a whole or as an extrasomatic element, such as, for example, the dependence of the axon on trophic influence from the soma.[49] It can be stated that at higher levels of the central nervous system in young mammals, degeneration of fibers with their synapses results in a remarkable growth of adjacent intact fibers, so that there is reoccupation of vacated synaptic sites. It can be postulated[49] that the trophic signal is provided, at least in part, by the degenerating fibers, in a manner that has been well established in the peripheral nervous system. This trophic influence may be mediated by glial cells, just as occurs with Schwann cells in the peripheral nervous system. Doubtless the vacated synaptic sites also contribute attractive influences, thus ensuring their reoccupation, even if it be by synaptic terminals that in the original growth avoid these sites because of their different specificities. These regenerations in young animals occur only for short distances—a few hundred microns at the most. In adult mammals, regeneration has been observed only in special situations where the new growth would extend for tens of microns at the most.

## Genesis of Neuronal Locus Specificity

The location of nerve cells and the trajectories of nerve fibers are remarkably similar in all individuals of the same species. Many observations over the past century have confirmed this invariance of neural circuitry in many different species of animals, and there are also many examples showing that the assembly of components of the nervous system in the embryo proceeds along orderly, nonrandom lines. These observations led to the introduction of the term "neuronal specificity" to denote the singular features which distinguish one type of nerve cell from another. This term has been applied uncritically to at least two major differences between various types of neurons, based, first, upon the differences in their positions and, second, on their structural differences. Therefore, a clear distinction must be made[50] between neuronal phenotypic specificity, when referring to the morphological differences between various types of nerve cells, and neuronal locus specificity, when referring to the unique and specific position-dependent properties that nerve cells exhibit.

Let us consider the example of the visual system, in which neuronal phenotypic specificity is displayed by the various types of nerve cells that are found, for example, in the retina. Retinal ganglion cells differ in their morphology from retinal bipolar, amacrine, horizontal, or photoreceptor cells. Not only are cells of different types clearly found in different positions, but cells of the same neuronal phenotype may also be arrayed in a spatial pattern and have significant position-dependent differences. These position-dependent differences are well exhibited by the retinal ganglion cells, each of which sends its axon to a specific locus in the brain, so that their spatial order is projected as a retinotopic map into the visual centers of the brain.

To identify this type of specificity clearly, the term "neuronal locus specificity" has been introduced[50] and defined as the property of the individual retinal ganglion cell which predisposes its axon to terminate and synapse at a particular locus in the retinotectal map. In normal frogs, ganglion cells at a particular retinal position always carried a particular place in the optic tectum. Thus, the spatial pattern of the individual retinal axon terminals in the tectum reduplicates the spatial relationship of the corresponding retinal ganglion cells.

In order to study the developmental origins of the position-dependent properties in the retinal ganglion cells, the experimental strategies involved were to reposition one eye of the embryo of the clawed frog, *Xenopus laevis*, at various stages of development before the eye had started forming nervous connection with the brain.[50] The eye was either inverted *in situ*, within the ocular orbit, or it was transplanted to another position on the body, or it was explanted into tissue culture for some time. In all cases the eye was then reintroduced to the orbit in order to assay the resulting retinotectal projection.

The first type of experiment was to excise the eye rudiment at various embryonic stages and then simply to reimplant it in different orientations in its own ocular orbit, or to transplant it from a donor to a carrier at the same stage of development.[50] When the eye was rotated at embryonic stage 28, it gave rise to a normal retinotectal projection when mapped after metamorphosis. However, when an eye rotation occurred a few hours later, at embryonic stage 30, a

retinotectal projection developed that was inverted in the anteroposterior axis of the retina but normal in its dorsoventral axis. Slightly
later, at stage 31, an eye rotation led to complete inversion of the
retinotectal map. These experiments show that there is a change in
the embryo during a 5- to 10-hour *critical period* between stages 28 to
32, during which there is a change in response of the retinotectal
system to 180° rotation of the eye. These experiments permit three
main conclusions.

1. The set of locus specificities that ultimately develop in an eye
rotated before the critical period is spatially organized, in accordance
with the postoperative positions of the retinal cells, and does not
depend on the cells' preoperative positions.

2. In an eye rotated after the critical period, the set of locus
specificities that ultimately develops is determined by the preoperative positions of the retinal cells and does not pay heed to their postoperative positions. By rotating an eye in the middle of the critical
period, at about stage 30, the set of position-dependent properties
that develops in the retina has an anteroposterior component appropriate to the eye's preoperative position, but the dorsoventral component is appropriate to its postoperative positions. Thus, the position dependence of locus specificity has two axial components, one
related to the ganglion cell's position in the anteroposterior axis of the
retina and the other to the cell's position in the dorsoventral axis of
the retina.

3. Although the eye at stage 32 contains only a few hundred
retinal ganglion cells, and retinal axons have not yet grown out of the
eye, rotation of the eye at stage 32 results in corresponding rotation of
the entire retinotectal map of the adult eye, which contains about
500,000 retinal ganglion cells. Apparently a developmental program
has been established in the embryo that affects the entire set of specificities that will arise in the adult retina.

Two main limitations of such an experimental analysis have been
emphasized.[50] The first relates to the long delay between the experimental manipulation of the embryonic eye rudiment and the final
mapping of the ultimate set of locus specificities as they are expressed
in the retinotectal map of the adult frog. During the intervening
period, there is a protracted process of generation of retinal and tectal
cells, the growth of the retinal nerve fibers into the brain, and the
formation of retinotectal connections. Although the experimental
analysis shows that the complete set of locus specificities that finally
develops in the adult retina is specified (or determined) during a short
critical period in the embryo at stages 29 to 32, one cannot infer

anything about the cellular mechanisms or the histogenetic programs that give rise to the spatial deployment of locus specificities and their expression during morphogenesis of the retinotectal map. Another limitation relates to the information that can be obtained from the map itself. The map provides no more than relative information about the disposition of the position-dependent properties across the retinal cell population. The map does not show which locus specificities are present; it only shows the order of their spatial deployment in the retinal cell population. Thus, it is extremely risky to try to correlate discontinuities or continuities in the retinal fiber projection to the tectum with a history of surgical manipulation of the retinotectal system. At best, the map permits a correlation between the direction in which these locus specificities are deployed in the eye and the previous developmental history of the eye's orientation with respect to the body. Thus, the main limitation of the technique is that in an experimental eye it cannot assay the range of the position-dependent properties or tell whether the set of properties is complete, reduced, or augmented. However, it can provide information about the relative order with which the set of position-dependent properties has been spatially deployed in the retina. A recently invented technique makes it possible to assay the range and the set of properties of the ganglion cells in an experimentally manipulated eye.[50] An experimental eye was grafted in the same eye socket as a normal eye at about embryonic stage 32.[50] Both the normal and the experimental eye project their optic nerve fibers into the same tectum. The normal eye serves as a standard, since it is presumed to contain a full and normal set of position-dependent properties, and the mode of intercalation of the fibers from the two eyes into that part of the optic tectum that they share can provide quantitative information about the range and the set of properties of the retinal ganglion cells in the experimental eye. Such an assay has shown, for example, that eyes inverted *in situ* contain a full set of position-dependent properties.

As the retinotectal mapping assay provides only limited information about the properties of retinal ganglion cells and none about the connections between retinal and tectal cells, it is necessary to use postsynaptic recording techniques as a crucial step toward an understanding of the expression of neuronal locus specificity.[50] Without information about the transactions between the optic nerve fiber tips and the tectal cells which result in the formation of functional retinotectal synapses, the understanding of the expression of cellular mechanisms that determine how nerve fibers reach the proper places and how nerve fiber tips discriminate amongst potential postsynaptic cells cannot be understood.

Returning again to the strategies used for studying the genesis of position-dependent properties in the retinal ganglion cells, first one has to show that the position-dependent properties are truly dependent upon position in the retinal cell population and are not merely an expression of some temporal order in development, such as the order of origin of retinal ganglion cells, the order of their differentiation, and the timing of the arrival of the retinal axons in the tectum.[50] That the latter was not the case was shown in experiments in which eyes were either grafted from stage 32 embryos to the body wall for 30 days, or explanted into tissue culture for 10 days before being reimplanted in the orbit and allowed to form connections with the brain. These eyes formed normal retinotopically organized projections with orientations appropriate to the positions of the eyes in the original donors.[50] When such an eye was allowed to compete with a normal eye during innervation of a single tectum, both projected in register with no parts of the tectum being unshared. These experiments showed that the expression of locus specificities does not depend upon the sequential order of arrival of nerve fibers from the eye into the tectum. It was then necessary to discover whether the position dependence was derived from the position of the ganglion cell within the retinal field or was derived from the position of the eye on the body surface. For this purpose a stage 28 eye rudiment was transplanted onto the body wall of a stage 28 intermediate host in various orientations. After the eye had developed to mid-stage 30, it was reintroduced to the orbit of a stage 39 carrier embryo and allowed to form connections with the brain. The resulting retinotectal projections were normally organized and were oriented in a position appropriate to the orientation of the eye on the body wall. When such an eye was allowed to compete with a normal eye in the innervation of a single tectum, both eyes projected to the entire tectum with no area left unshared. The set of specificities in the experimental eye, which had spent its critical period on the body wall, was indistinguishable from the set of specificities in the normal eye. Therefore, the position-dependence of locus specificity is derived not from some absolute position on the body surface but from the relative positions of the ganglion cells within the retinal field.[50] Thus, the eye is not directly instructed by the embryo about its absolute position on the body but merely uses the body as a source of positional cues for establishing properly aligned axes of its own.

Clearly the axial cues are not unique to the tissues surrounding the eye but are also available to an eye on the side of the body. This raises the possibility that the same cues may be used to align the axes of other organ rudiments such as the limb or the ears. Like the eye, these organ rudiments also acquire axial cues from the surrounding tissues, which then participate in aligning the axes of the organ rudiment with the axes of the embryo.[50]

The next question that arises is whether the transition from the unspecified to the specified state involved stable and irreversible changes in the retinal cells or whether, in contrast, the change in the state of the system reflects changes in the extraocular conditions in the embryo. An example of the latter change might be the disappearance of axial cues required to reorganize retinal axes after eye rotation or transplantation of the eye. To test whether the transition occurring at stages 29–31 involves irreversible changes in the eye or in the embryo, right eyes at stages 31–32 were back-grafted into the enucleated right orbits of stage 28 embryos.[50] When the eye was back-grafted in an inverted position, the pattern of retinotectal connections that developed was retinotopically organized but inverted in both axes of the tectum. When the back-grafted eye was introduced into a stage 28 orbit in normal orientation, a normally oriented retinotectal projection invariably developed. These results show that the back-grafting procedures themselves did not alter the pattern of retinotectal connectivity and that a back-grafted eye at stages 31–32 developed locus specificities from positional information obtained in the original donor orbit and failed to show the influence of its new position in the host orbit. The eye at stages 31–32 is thus refractory to the same conditions which are capable of providing positional information to an unspecified eye in a stage 28 animal. In addition, the independent stability of the anteroposterior and dorsoventral axial components of locus specificity by back-grafting left eyes at stages 31–32 into stage 28 right orbits has been tested. Such a transfer inverts one retinal axis only. In these instances the retinotectal projections developing from the back-grafted eye were inverted in one axis and normal in the other. The independent stability of the two axial components of locus specificity was confirmed in several cases in which eyes were back-grafted at stage 30, intermediate specification in the two axes. Thus, anteroposterior inversion of the retinotectal map resulted from back-grafting stage 30 right eyes, in an inverted orientation, into stage 28 right orbits.

These experiments provide strong evidence that the transition from the unspecified to the specified state involves stable and irreversible changes in the eyes between stages 29 and 32, but whether irreversible changes also occur in the embryo remains unknown. That the axial cues persist until stage 39, and are available to a stage 28 eye implanted into a stage 39 embryo, was shown in the following experiment. A stage 23 eye was transplanted in rotated position into a stage 28 host, allowed to traverse the critical period of that host, and then reimplanted into a stage 39 carrier embryo.[50] The reimplantation was done before the eye itself had reached its critical period. Such eyes always developed normal retinotectal projections regardless of their orientation in the final carrier embryo. These experiments show that the trigger mechanism for the transition from unspecified to specified state is not the sudden disappearance at stage 28 of the extraocular axial cues, which presumably participate in organizing the retinal axes and aligning them with the major axes of the body. Rather, the trigger mechanisms are within the eye itself. That the transition during the critical period was not due to extraocular agents that transiently arise between stages 28 and 32 was shown in the following experiment: Unspecified eyes before stage 28 were transferred directly to carrier embryos between stages 32 and 39. In all cases, regardless of the orientation of the eye in the final carrier, a normally oriented retinotectal projection developed. Thus, a normal pattern of retinotectal projections developed from an eye which had never been in contact with an embryo at the critical stages of 28–32. These experiments eliminate a classical induction model of specification in which extraocular agents arise transiently during the critical period and instruct or specify an unspecified retinal cell population.

The specification process does not involve instructive action on the part of the embryo, but merely reflects a transition from a reversible to an irreversible state. Eye rudiments from embryos at stages 22–25 were cultured *in vitro* for periods of 24 days, during which time the eyes developed the features of stage 38 or 39 normal eyes. These mature explants were then reimplanted into stage 39 carrier embryos.[50] In these cases the resulting retinotectal maps were normal when the eye was in its normal orientation in the final carrier, but when the eye was rotated the map was rotated to the same extent, showing that the locus specificities in the retinal ganglion cell population were derived from the original donor and were not altered in the final carrier. These experiments show that specification occurs in tissue culture in total isolation from the embryo, and that the eye, as early as stage 22, contains a set of reference axes properly aligned with the axes of the stage 22 embryo. The embryonic eye in the unspecified state before stage 28 possesses a stable but reversible set of axes. Whether these axes are physically the same as those present after stage 31 remains to be determined. However, it is certain that the specification process merely consists of a transition from a reversible but stable set of axes in the unspecified retina to a stable but irreversible set of retinal axes in the specified retina.

As specification occurs in isolation from the embryo, it requires no instructive action from the tissues outside the eye. The change in the eye that renders the retinal cells refractory to the continuing influence of axial cues from the surrounding cues remains to be considered. The foregoing discussion has defined the changes of state undergone by the retinal cell population in the development of definitive locus specificities, but it has left open the question of cellular mechanisms. One would like to elucidate the cellular mechanisms of (1) deployment of position-dependent properties in the retinal cell population, (2) substitution of new retinal axes after eye inversion, and (3) expression of the properties as locus specificities during the morphogenesis of the retinotectal projection. One limitation of the experiments that have been described is that, while they are able to define the state of the total retinal cell population in the intact eye, they do not show the state of individual retinal cells. An additional problem arises from the fact that at stages 29–30 the retina contains about 500 cells, but the locus specificities are determined not only for the cells then present but for all the retinal ganglion cells (about 500,000) that will ultimately develop in the adult eye.[50] Because one operates on the embryos but only assays the locus specificities in the adult eye, it is not known when the final set of locus specificities develops. The retina grows by addition of new cells at its margin, where a population of retinal stem cells persists throughout embryonic and larval development.[50] The question then arises as to whether the nerve cells acquire specificity from the stem cells as a result of cell lineage, or whether, in addition or exclusively, they acquire specificity by interacting with other cells. If the former is true, how do the stem cells pass on the specificity to their progeny? If the latter, do the newborn nerve cells interact with other cells within the eye or with cells outside the eye, or with both?

Another problem arises because nerve cells die during normal retinal histogenesis.[50] How is the pattern of deployment of position-dependent properties in the retinal cell population affected by cell death? Other unanswered questions are: Why do some neurons die and others survive; is cell death random or selective; do ganglion cells die before, during, or after the arrival of their axons in the tectum; and does ganglion cell death play any part in sorting of retinal fibers in the tectum?

At present, our only clues to a solution of these problems are certain correlations between retinal histogenesis and the genesis of locus specificities in the retinal ganglion cell population. First, the retinal ganglion cells born in the central region of the retina cease DNA synthesis at stages 29–31,[50] and it seems likely that changes occur in their developmental programs accompanying their withdrawal which initiates the transition from the unspecified to the specified state. One may, with some confidence, propose the hypothesis that in many different neuronal populations the change from the unspecified to the specified state occurs as nerve cells become postmitotic; that is, at the time of their birth.[50] One cause of the change of state occurring at that stage of the nerve cell's development may be that the postmitotic cell uncouples from the stem cell and thus loses the capacity to interact with the stem cell population. Apparently, developing nerve cells, when grafted to another position, can acquire new position-dependent properties only if they can couple functionally with neighboring cells. The locus specificities of the grafted cells then develop in accord with their new positions. However, if the graft itself cannot communicate with neighboring cells, its locus specificities are derived from the original position before grafting. Intercellular communication of the type envisaged here appears to depend upon specialized cellular junctions[51] that provide for the movement of ions and small molecules between the cells of the embryo.[51,52] Ubiquitous intercellular junctions of the gap junction variety have been found between the cells of the embryonic retina of *Xenopus* before stage 29 (while the retina is composed of mitotically active stem cells), but these disappear from the retinal cells in the center of the retina during the critical stages 29–32 (as the first retinal nerve cells, destined to become ganglion cells, withdraw from the mitotic cycle), and at later stages persist only between the cells at the retinal margins (which remain mitotically active). Some experiments[51] have demonstrated the passage of fluorescent-labeled molecules between cells of the embryonic retina before stage 30, but have shown that such intercellular communication is impeded or absent at later stages of development of the retina. Therefore, it seems that the change of state at the critical period involves withdrawal of retinal cells from the mitotic cycle and concomitant uncoupling of the postmitotic nerve cells from the rest of the retinal cell population. After its uncoupling from the retinal population, the nerve cell is characterized by a stable and irreversible position-dependent property that results in the development of definitive locus specificity. Development of this position-dependent specificity enables the young nerve cell to extend its outgrowing axon on the proper pathway, predisposing that axon to terminate at a specific position in the brain and to form selective synaptic connections with particular nerve cells. Therefore, the position-dependence of the nerve cell is expressed at several stages of its developmental program. How this control of the nerve cell's position occurs in terms of cellular mechanisms remains an unsolved problem of developmental neurobiology.

## FUNCTIONS OF CHOLINERGIC INPUT

The cholinergic input to the superior cervical ganglion has been shown to modulate the differentiation of sympathetic neurons. A significant increase in TH activity occurs after the development of cholinergic synapses in the ganglia; furthermore, decentralization of the ganglion to prevent its cholinergic innervation significantly reduces developmental increases in TH activity.[53,54] Treatment with drugs that block the nicotinic cholinergic receptor at the critical time also reduces the subsequent increase in TH activity, thus demonstrating that acetylcholine liberated by the presynaptic terminals is the chemical cue that promotes the differentiation of postsynaptic neurons.[55] The cholinergic input regulates not only enzyme levels in the perikarya but also the process of target organ innervation.[56]

## NEURONAL PLASTICITY

The nervous system of all animals is an ordered network of interconnecting neurons and supporting elements. It is unknown how elements of the central and peripheral nervous system form and maintain their networks or circuits, which do not appear to vary within the same animal species. This basic behavioral invariance suggests that at least some reflex arcs are predetermined; however, it is not yet known which connections within any given nervous system are under genetic influences.

It is unlikely that the genetic materials within any given nervous system are sufficiently ample to code for all neural interconnections within the same system.[57] Yet some behavioral patterns and reflex arcs appear definitely under genetic controls and therefore inheritable.[58-62] Whether these behavioral patterns and reflex arcs are controlled to such an extent that each connection made is predetermined is not known. It has been postulated[63,64] that the type I nerve cells (long-fiber neurons) may form a sort of neural "skeleton," upon which the short-fiber type II neurons are hung. Since it is most important for the skeleton to be properly interconnected, these neurons may be predetermined from their birth. The type II nerve cells, which are produced in redundant quantity later in embryonic life, are thought to respond to environmental factors, allowing the animal to adapt to situations unique to its daily existence.

Since some form of exact interconnection exists between restricted parts of a given nervous system, there is an obvious importance in determining what controls are exercised over the formation of such connections and whether these controls can be influenced or molded by experimental design. Neural connections, once broken, often display the ability to reform themselves in such a manner as to restore the original function of the broken circuit. Such abilities are greater in invertebrates, embryos, and lower vertebrates, but diminish in scope with increasing age and phylogeny. These repair processes often take place over considerable distances, such as optic nerve regeneration in fish and amphibians, or peripheral regeneration of sensory and motor neurons throughout most classes of invertebrates and vertebrates. Time of recovery from such lesions depends on the class of animal species under examination (fish and amphibians repair more rapidly than mammals), age of the animal (embryonic, larval, and neonatal animals repair more rapidly than adults), and site and extent of the experimental lesions (CNS lesions generally repair less rapidly or fully than peripheral axotomies).[65]

## Peripheral Plasticity

Section of peripheral nerve trunks, either motor or sensory, frequently results in regeneration of the sectioned fibers and a return of function in the peripheral end organ. Depending upon the species of animal used, the degree of successful regeneration varies. When regeneration does occur, however, generally the old nerve trunks and original innervation sites are preferentially reinnervated.[66-69]

The evidence for specific regeneration of peripheral motor nerve supplies in the salamander has been reported.[70] Crossing the nerve supplies to forelimb extensor and flexor muscles did not prevent the experimental forelimb from moving in a normal fashion. Gross dissection revealed that the nerve fibers remained crossed, and their functional connections with the inappropriate muscles were determined electrophysiologically. Stimulation of the nerve trunks further proximal to the crossover showed, however, that some of the original nerve supplies to the correct muscles remained intact. This shows that a few of the original fibers somehow found their way back to their original muscles, thus producing normal reflex movements.

Hyperinnervation of fish muscles and selective reinnervation of muscles in amphibian limbs occurs.[17] Gross rerouting of large nerve trunks failed to produce maladaptive limb movements following reinnervation. Such gross rerouting made it impossible for the regenerating nerve fibers not to make contact with the wrong set of muscles. Profuse sprouting by the severed nerves, on the other hand, increased the possibilities of correct muscle reconnection by the appropriate motorneurons. Section of a spinal nerve trunk in the salamander results in the expansion of adjacent motorneuron receptive fields.[71,72] Marginal extension of adjacent nerve fibers into denervated muscles began 2–3 days after nerve trunk section, with maximum spread of adjacent innervation occurring during the following 2 weeks. Upon reinnervation of the muscle by its original nerve trunk, adjacent, expanded nerve supplies failed to elicit motor responses. A second section of the normal nerve trunk led to a rapid spread of cross-innervation which occurred within 3 days of section.

Such a rapid spread of responsiveness could not be accounted for by mere collateral sprouting, but was believed to be due to a reemergence of function in preexisting, but silent, nerve terminals. Cutting adjacent nerve trunks confirmed this overlap in adjacent receptive fields. Following section of an adjacent nerve trunk, degenerating nerve terminals within the experimental muscle were found in ratio of approximately three normal terminals to one degenerating terminal. The reverse of this ratio was observed within the marginal areas of the experimental muscle when its nerve supplies were sectioned. The conclusion drawn from these experiments was that competitive interactions occur between nerve fibers for control of muscles, and that multiple innervation between correct and incorrect nerve terminals can occur within the same muscle without interfering with normal function of the muscle.

One need not perform an axotomy on a neuron in order to produce collateral sprouting of its neighbor terminals. Interruption of axoplasmic flow in an axon often leads to cell body changes resembling chromatolysis[73] and to a reduction in strength of synaptic transmission.[74] Direct application of colchicine solutions to spinal nerves of the salamander *Ambystoma tigrinum* results in a cessation of axoplasmic flow in the experimental nerve and an expansion of sensory and motor receptive fields of adjacent spinal nerves in the periphery.[75] The expansion of hindlimb nerves 15 and 17 occurred in identical manners when nerve 16 was either sectioned or treated with the drug. Application of colchicine to either nerve 15 or 17 prevented peripheral sprouting even after nerve 16 had been sectioned. It was concluded that trophic factors occur within the neurons themselves which regulate the extent of peripheral receptive fields. Absence of these factors results in a situation paralleling axotomy and a subsequent invasion of the affected neuron's receptive fields by its neighbors. The application of colchicine to neighboring nerve fibers prevents them from sprouting, presumably also because of a lack of materials necessary for such growth to occur.

### Central Plasticity

Plastic changes also occur within the CNS in response to localized damage. The extent of such repair processes, either through effective regeneration or collateral sprouting of intact adjacent fibers, is greatly reduced when compared to the rather extensive peripheral nerve repair processes. Like peripheral nerve repair process, central regeneration and collateral sprouting are influenced by ontogenetic and phylogenetic considerations, the least amount occurring in adult mammals. That some central neurons fail to regenerate or be replaced following cell death has been noted for many years[76,77]; yet limited plastic changes in neurons adjacent to a central lesion have been noted as well.[76]

Some studies suggest that one group of neurons, the monoamine-containing neurons of the mammalian CNS, may undergo a more extensive regenerative process than previously believed.[77,78] One example of these plastic changes among autonomic neurons has been noted in the selective reinnervation of the cat superior cervical ganglion.[80] When a selected input of preganglionic nerve fibers to the superior cervical ganglion of the cat is removed, adjacent thoracic nerve roots invade and functionally innervate the ganglion. Upon reinnervation by the original preganglionic nerve supplies, an apparent specific displacement of the anomalous preganglionic nerve endings from the postganglionic cells by the original nerve fibers occurs.[81]

Several dorsal roots in the cat resulted in extensive ipsilateral collateral sprouting among the unaffected, adjacent dorsal root fibers.[82] This was demonstrated histologically by cutting dorsal roots on either side (singly or together) of an initial dorsal root lesion and noting the increase in axonal debris within the dorsal spinal cord. Section of a single dorsal root resulted in a generalized and diffuse sprouting in the adjacent, ipsilateral dorsal root fibers. Lesions within the corticospinal tract also produced collateral sprouting within the spinal cord, but within localized areas of the bulbar pyramid. Two forms of collateral sprouting in adult cats were noted: (1) a generalized collateral axonal sprouting of the intraspinal dorsal root projection; and (2) a limited sprouting to only one side of the spinal cord and within only selected regions of the nucleus gracilis. Diffuse central sprouting has been reported by numerous investigators[76,83,84] and may merely represent a generalized response to axotomy. However, limited specific axonal sprouting has been reported,[85-89] and has attracted attention as a possible means of examining the specificity with which neurons form connections within the CNS.

An exciting possibility now exists that, given the opportunity, peripheral and central nerve fibers undergoing collateral sprouting or regeneration would exhibit selective reinnervation or readjustments.[65] Such specific reconnections have been examined in both *in vivo* and *in vitro* systems.[90-94] Preferential associations occur between structures connected normally in the intact animal and their normal nerve supplies. Sympathetic ganglia explants from rat were cultured between two equally proliferating target tissues. A preferential variation in the number of fibers growing in the direction of the target explants was observed to occur. It has been proposed that the extent of direct nerve fiber growth to a tissue type might be related to the density of its normal potential for sympathetic innervation. Thus, explants of atria and vas deferens, which normally receive prodigious amounts of sympathetic innervation, attracted more fibers than did kidney medulla, uterus, and ureter, which are sympathetic-poor tissues. Similar selectivity of reconnections has been observed between cockroach metathoracic ganglia and limb segments, rat superior cervical dissociated ganglia and spinal cord, optic tectum and neural retina, and leech touch cells in adjacent segmental ganglia.[67]

Limited central sprouting and specific plastic changes occur within the rat CNS following localized lesions of a given nerve tract.[93] Electrolytic lesions were made in the medial forebrain bundle or the fimbria of the rat. The septal nuclear complex was then studied under the electron microscope. The septal nuclear complex in the rat receives afferent fibers from the two sources mentioned above, with the fimbrial fibers synapsing selectively on dendrites and the hypothalamic fibers terminating on both septal dendrites and perikarya. Long-term lesion studies were conducted on one of the two inputs to the septal nuclei, followed by short-term lesion studies to examine the distribution patterns of the remaining afferents. The findings revealed that when long-term fimbrial lesions had been made, an unusually dense pattern of dendrite synapses occurred from the remaining, intact hypothalamic afferents. Chronic hypothalamic lesions resulted in fimbrial afferents forming synapses on cell bodies. It was concluded that there is a reoccupation of the deafferented sites by local intact terminals, even though the connections made are not the correct connections.

Section of facial nerve peripherally causes displacement of synaptic boutons from the nerve cell body during the period of chromatolysis and recovery.[95] Hemisection of the spinal cord of rat, monkey, and man also results in considerable axonal sprouting, including the formation of new synapses.[96] Axonal sprouting of sensory endings within the cat spinal cord was the subject of a series of sophisticated experiments by Eccles et al.[85] In the cat, highly specific monosynaptic reflex arcs exist between the sensory stretch afferents and the motorneurons of their muscle of origin and its synergists. It was felt that cross-union of motor nerve trunks in a neonatal mammal might possibly result in central readjustments of the stretch neuron synapses upon various motorneurons which could be measured electrophysiologically. These investigations were designed, then, to test the theory of myotypic modulation in a mammal on what appeared to be an optimal reflex circuit. In these experiments, the nerve trunks to the peroneal muscles and medial gastrocnemius or the plantaris and lateral gastrocnemius muscle were cross-united and allowed to reinnervate their new end-organs. Several months later, EPSP recordings were carried out on spinal motorneurons. Normally the largest EPSPs are recorded from afferents innervating the same or synergistic muscle groups. Changes in the character of the EPSPs were detected in several individual cases, but the hoped-for, clear-cut changes in synapses could not be determined universally. Following the operations there was a significant decrease in the number of afferent fibers left in the nerve trunks as well as massive central degeneration of terminals also in dogs.[97] The changes recorded by these investigators could not, then, be solely attributed to a simple plastic switching of synapses. There were, however, several instances where changes in connections were felt to be occurring and not masked by the morphological changes observed.

It is unknown if neurons are predetermined from their birth to connect with their ultimate peripheral and central end organs, or whether some degree of peripheral influence can be exerted on the young neuron so as to direct the development of the finished neural circuit in the centers. It has been shown that peripheral manipulations can influence the development of central connectivity in frogs and cats. It is possible that such influences are frequently utilized to mold other connective patterns within a given nervous system.[65]

It has also been postulated that neurons might be biased towards the type of reflex associations they will make in the CNS either from birth or from an initial contact with embryonic end organs. Thus, nerve fibers innervating early embryonic back skin in the frog may be biased to establish reflex pathways mediating hindlimb wiping reflex responses. Being biased through embryonic contact with the periphery, however, does not mean that these cutaneous neurons would be incapable of changing or modifying their characteristics. One must also consider that even though a neuron might acquire certain affinities from birth or from an early contact with the periphery, the possibility exists that final confirmation of these affinities might not be effected until some dramatic finalizing stage of development has been reached, perhaps when the neuron becomes part of a stable circuit. If the nerve cell processes an either/or capability prior to the formation of stable associations, the type of cell it will become and the type of connections it may make in the centers would depend quite heavily upon what types of confirming information it received from the periphery.[65]

No single, simple conclusion can be drawn from the various investigations that would answer the question of how nerve cells develop their patterns of interconnectivity, either initially during embryogenesis or during regeneration at later stages of development. In all of the experiments designed to test the bases for specific connectivity of neural tissues *in vitro*, the strongest indications are that some form of intercellular recognition exists associated with the cell membrane.[65] As differentiation and maturation of nerve cells are accompanied by greater structural complexity, one might expect membrane systems to undergo structural changes that may be accompanied by alterations in their protein–lipid makeup subserving newly acquired functions. Whether selective neural interconnections are indeed modifiable biochemical phenomena remains to be determined.

## REFERENCES

1. Weston, J. A., The migration and differentiation of neural crest cells, *Adv. Morphog.* 8:41–114, 1970.

2. Le Douarin, N., and Le Lievre, C., Embryologie Experimentale. Démonstration de l'origine neurale des cellules à calcitonine du corps ultimobranchial chez l'embryon de poulet, C. R. Acad. Sci. (Paris) Ser. D. **270**:2857–2860, 1970.

3. Johnston, M. C., Bhakdinaronk, A., and Reid, Y. C., An expanded role of the neural crest in oral and pharyngeal development, in: Fourth Symposium on Oral Sensation and Perception (J. F. Bosma, ed.), pp. 37–52, USGPO, 1973.

4. Noden, D., An analysis of the migratory behavior of avian cephalic neural crest cells, Dev. Biol. **42**:106–130, 1975.

5. Weston, J. A., A radioautographic analysis of the migration and localization of trunk neural crest cells in the chick, Dev. Biol. **6**:279–310, 1963.

6. Le Douarin, N. M., Renaud, D., Teillet, M. A., and Le Douarin, G. H., Cholinergic differentiation of presumptive adrenergic neuroblasts in interspecific chimeras after heterotopic transplantations, Proc. Natl. Acad. Sci. U.S.A. **72**:728–732, 1975.

7. Le Douarin, N., A biological cell labeling technique and its use in experimental embryology, Devel. Biol. **30**:217–222, 1973.

8. Le Douarin, N., and Barq, G., Embryologie Expérimentale. Sur l'utilisation des cellules de la Caille japonaise comme (margueurs biologigue) en endryologie experimentale, C.R. Acad. Sci. (Paris) Ser. D. **269**:1543–1546, 1969.

9. Dorris, F., The production of pigment in vitro by chick neural crest, Wilhelm Roux' Arch. Entwicklungsmech. Org. **138**:323–335, 1938.

10. Cohen, A. M., and Konigsberg, I. R., A clonal approach to the problem of neural crest determination, Dev. Biol. **46**:262–280, 1975.

11. Maxwell, C., Cell cycle changes during neural crest cell differentiation in vitro, Dev. Biol. **49**:66–79, 1976.

12. Drews, U., Kocher-Becker, U., and Drews, U., Die Induktion von Kiemenknorpel aus Kopfneuralleistenmaterial durch präsumptiven Kiemendarm in der Gewebekultur und das Bewegungsverhalten der Zellen während ihrer Entwicklung zu Knorpel, Wilhelm Roux' Arch. Entwick Tungsmech, Org. **171**:17–37, 1972.

13. Newsome, D. A., In vitro stimulation of cartilage in embryonic chick neural crest cells by products of retinal pigmented epithelium, Dev. Biol. **49**:496–507, 1976.

14. Cohen, A. M., Factors directing the expression of sympathetic nerve traits in cells of neural crest origin, J. Exp. Zool. **179**:167–182, 1972.

15. Norr, S., In vitro analysis of sympathetic neuron differentiation from chick neural crest cells, Dev. Biol. **34**:16–38, 1973.

16. Okun, L. M., Isolated dorsal root ganglion neurons in culture: cytological maturation and extension of electrically active processes, J. Neurobiol. **3**:111–151, 1972.

17. Petersen, E. R., and Murray, M. R., Myelin sheath formation in cultures of avian spinal ganglia, Am. J. Anat. **96**:319–355, 1955.

18. Cowell, L., and Weston, J. A., An analysis of melanogenesis in cultured chick embryo spinal ganglia, Dev. Biol. **22**:670–697, 1970.

19. Weston, J. A., Neural crest cell migration and differentiation, in: Cellular Aspects of Neural Growth and Differentiation, (D. Pease, ed.), UCLA Forum in Medical Sciences, No. 14, pp. 1–22, University of California Press, Los Angeles, 1971.

20. Newgreen, D. F., and Jones, R. O., Differentiation in vitro of sympathetic cells from chick embryo sensory ganglia, J. Embryol. Exp. Morphol. **33**:43–56, 1975.

21. Patterson, P. H., Reichardt, L. F., and Chun, L. Y., Biochemical studies on the development of primary sympathetic neurons in cell culture, Cold Spring Harbor Symp. Quant. Biol. **40**:389–398, 1975.

22. Burnham, P., Raiborn, C., and Varon, S., Replacement of nerve growth factor by

ganglionic non-neuronal cells for the survival *in vitro* of disassociated ganglionic neurons, *Proc. Natl. Acad. Sci. U.S.A.* **69**:3556–3560, 1972.

23. Weston, J. A., Pintal, J. E., Derby, M. A., and Nichols, D. H., The morphogenesis of spinal ganglia from neural crest cells, in: *Progress in Clinical and Biological Research*, Vol. 15 (Z. Hall, R. Kelly, and C. F. Fox, eds.), pp. 217–226, Alan R. Liss, New York, 1977.

24. Pratt, R. M., Larsen, M. A., and Johnston, M. C.; Migration of cranial neural crest cells in a cell-free hyaluronate-rich matrix, *Dev. Biol.* **44**:298–305, 1975.

25. Cohen, A. M., and Hay, E. D., Secretion of collagen by embryonic neuoepithelium at the time of spinal cord-somite interaction, *Dev. Biol.* **26**:578–605, 1971.

26. Scott, J. E., and Dorling, J., Differential staining of acid glycosmino-glycans (mucopolysaccharides) by Alcian blue in salt solutions, *Histochemie* **5**:221–233, 1965.

27. Pollister, A. W., Swift, H., and Rasch, E., Microphotometry with visible light, in *Physical Techniques in Biological Research*, Vol. 3c (A. W. Pollister, ed.), pp. 201–251, 1969.

28. Pintar, J. E., Derby, M. A., and Weston, J. A., The interaction of trunk neural crest cells with glycosaminoglycans, *J. Cell. Biol.* **70**:371a, 1976.

29. Weston, J. A. and Butler, S. L., Temporal factors affecting localization of neural crest cells in the chicken embryo, *Dev. Biol.* **14**:246–266, 1966.

30. Toole, B. P., Hyaluronate turnover during chondrogenesis in the developing chick limb and axial skeleton, *Dev. Biol.***29**:321–329, 1972.

31. Black, I. B., Regulation of the growth and development of sympathetic neurons *in vivo*, in: *Progress in Clinical and Biological Research*, Vol. 15 (A. Hall, R. Kelly, and C. F. Fox, eds.), pp. 61–71, Alan R. Liss, New York, 1977.

32. Sperry, R. W., Chemoaffinity in the orderly growth of nerve fiber patterns and connections, *Proc. Natl. Acad. Sci. U.S.A.* **50**:703–710, 1963.

33. Roth, S., McGuire, E. J., and Roseman, S., An assay for intracellular adhesive specificity, *J. Cell Biol.* **51**:525–535, 1971.

34. Walther, B. T., Ohman, R., and Roseman, S., A quantitative assay for intercellular adhesion, *Proc. Nat. Acad. Sci. U.S.A.* **70**:1569–1573, 1973.

35. Gottlieb, D. I., Merrell, R., Rock, K., Littman, D., Santala, R., and Blaser, L., Studies on cell recognition in the developing brain, in: *Progress in Clinical and Biological Research*, Vol. 15 (Z. Hall, R. Kelly, and C. F. Fox, eds.), pp. 139–146, Alan R. Liss, New York, 1977.

36. Holtfreter, J., Studien zur Ermittlung der Gestaltungsfaktoren in der Organenentwicklung der Amphibien, *Wilhelm Roux' Arch. Entwicklungmech. Org.* **139**:110–190, 1939.

37. Moscona, A. A., Analysis of cell recombinations in experimental synthesis of tissues *in vitro*, *J. Cell. Comp. Physiol.* **60**(Suppl)(1):65–80, 1962.

38. Trelstad, R. L., Hay, E. D., and Revel, J. P., Cell contact during early morphogenesis in the chick embryo, *Dev. Biol.* **16**:78–106, 1967.

39. Hausman, R. E., and Moscona, A. A., Purification and characterization of the retina-specific cell-aggregating factor, *Proc. Natl. Acad. Sci. U.S.A.* **72**:916–920, 1975.

40. Merrell, R., Gottlieb, D. I., and Glaser, L., Embryonal cell surface recognition. Extraction of an active plasma membrane component, *J. Biol. Chem.* **250**:5655–5659, 1975.

41. Balsamo, J., and Lilien, J., Functional identification of three components which

mediate tissue-type specific embryonic cell adhesion, *Nature (London)* **251**:522–524, 1974.

42. Thiery, J., Brackenburg, R., Rutishauser, U., and Edelman, G., Adhesion among neural cells of the chick embryo, in: *Progress in Clinical and Biological Research*, Vol. 15 (Z. Hall, R. Kelly, and C. Fox, eds.), pp. 199–206, Alan R. Liss, New York, 1977.

43. Rutishauser, U., Thiery, J. P., Brackenbury, R., Sela, B. A., and Edelman, G. M., Mechanisms of adhesion among cells from neural tissues of the chick embryo, *Proc. Natl. Acad. Sci. U.S.A.* **73**:577–581, 1976.

44. Cowan, W. M., Neuronal deaths as a regulative mechanism in the control of cell number in the nervous system, in: *Development and Aging in the Nervous System* (M. Rockstein, ed.), pp. 19–34, Academic Press, New York, 1973.

45. Hughes, A., A quantitative study of the development of the nerves in the hind limb of *Eleutherodactylus martinicensis*, *J. Embryol. Exp. Morphol.* **13**:9–34, 1965.

46. Prestige, M. C., The control of cell number in the lumbar ventral horns during the development of *Xenopus laevis* tadpoles, *J. Embryol. Exp. Morphol.* **18**:359–387, 1967.

47. Romanes, G. J., Motor localization and the effects of nerve injury on the ventral horn cells of the spinal cord, *J. Anat.* **80**:117–131, 1946.

48. Levi-Montalcini, R., The origin and development of the visceral system in the spinal cord of the duck embryo, *J. Morphol.* **86**:253–283, 1950.

49. Eccles, J. C., Trophic influences in the mammalian nervous system, in: *Development and Aging in the Nervous System* (M. Rockstein, ed.), pp. 89–99, Academic Press, New York, 1973.

50. Jacobson, M., Genesis of neuronal locus specificity, in: *Development and Aging in the Nervous System* (M. Rockstein, ed.), pp. 105–116, Academic Press, New York, 1973.

51. Lowenstein, W. R., Intercellular communication, *Sci. Amer.* **222**:78–86, 1970.

52. Sheridan, J. D., Electrophysiological evidence for low resistance intercellular functions in the early duck embryo, *J. Cell Biol.* **37**:650–659, 1968.

53. Black, J. B., Geen, S. C., Inhibition of the biochemical and morphological maturation of adrenergic neurons by nicotine receptor blockade, *J. Neurology* **22**:301–306, 1974.

54. Black, J. B., Hendry, J. A., Iversen, L. L., Effects of surgical decentralization and nerve growth factor on the maturation of adrenergic neuron in a mouse sympathetic ganglion, *J. Neurochem.* **19**:1367–1377, 1972.

55. Black, J. B., Hendry, J. A., Iversen, L. L., Trans-synaptic regulation of growth and development of adrenergic neurones in a mouse sympathetic ganglion, *Brain Res.* **34**:229–246, 1971.

56. Black, J. B., and Mytilineou, C., Trans-synaptic regulation of the development of end organ innervation by sympathetic neurons, *Brain Res.* **101**:503–521, 1976.

57. Horridge, G. A., *Interneurons: Their Origin, Action, Specificity, Growth and Plasticity*, Freeman, San Francisco, 1968.

58. Atwood, H. L., and Wiersma, C. A. G., Command interneurons in the crayfish central nervous system, *J. Exp. Biol.* **46**:249–261, 1967.

59. Bentley, D. R., Genetic control of an insect neuronal network, *Science* **174**:1139–1141, 1971.

60. Benzer, S., Genetic dissection of behavior, *Sci. Amer.* **229**:24–37, 1973.

61. Taub, E., Perella, P., and Barro, G., Behavioral development after forelimb deafferentation on day of birth in monkeys with and without blinding, *Science* **181**:959–960, 1973.

62. Threlkeld, S. F. H., Procwat, R. A., Abbott, K. S., and Yeung, A. D., Genetically

based behavior patterns in *Drosophila melanogaster*, *Nature (London)* **247**:232–244, 1974.

63. Jacobson, M., Development, specification, and diversification of neuronal connections, in: *The Neurosciences*, Second Study Program (F. O. Schmitt, ed.), pp. 116–129, Rockefeller University Press, New York, 1970.

64. Jacobson, M., A plentitude of neurons, in: *Aspects of Neurogenesis* (G. Gottlieb, ed.), pp. 151–166, Academic Press, New York, 1970.

65. Baker, R., Some comments on central and peripheral plastic changes in nerve connections, in: *Molecular and Functional Neurobiology* (W. H. Gispen, ed.), pp. 47–86, Elsevier, New York, 1976.

66. Speidel, C. C., Adjustments of nerve endings, *Harvey Lect.* **36**:126–158, 1941.

67. Baylor, D. A., and Nicholls, J. G., Patterns of regeneration between individual nerve cells in the central nervous system of the leech, *Nature (London)* **232**:268–269, 1971.

68. Bennett, M. R., McLachlan, E. M., and Taylor, R. S., The formation of synapses in reinnervated mammalian striated muscle, *J. Physiol. (London)* **233**:481–500, 1973.

69. Burgess, P. R., English, K. B., Korch, K. W., and Stensaas, L. J., Patterning in the regeneration of type I cutaneous receptors, *J. Physiol. (London)* **236**:57–82, 1974.

70. Grimm, L. M., An evaluation of myotypic respecification in axolotls, *J. Exp. Zool.* **178**:479–496, 1971.

71. Cass, D. T., Sutton, T. J., and Mark, R. F., Competition between nerves for functional connections with axolotl muscles, *Nature (London)* **243**:201–203, 1973.

72. Stirling, V., The effect of increasing the innervation field sizes of nerves on their reflex response time in salamanders, *J. Physiol. (London)* **229**:657–679, 1973.

73. Pilar, G., and Landmesser, L., Axotomy mimicked by localized colchicine application, *Science* **177**:1116–1118, 1972.

74. Perisic, M., and Cuenod, M., Synaptic transmission depressed by colchicine blockade of axoplasmic flow, *Science* **175**:1140–1142, 1972.

75. Aguilar, C. E., Bisby, M. A., Cooper, E., and Diamond, J., Evidence that axoplasmic transport of trophic factors is involved in the regulation of peripheral nerve fields in salamanders, *J. Physiol. (London)* **234**:449–464, 1973.

76. Ramon y Cajal, S., *Degeneration and Regeneration of the Nervous System* (trans. R. M. May), Hafner, New York, 1959.

77. Das, G. D., and Altman, J., Studies on the transplantation of developing neural tissue into the mammalian brain. I. Transplantation of cerebellar slabs into the cerebellum of neonate rats, *Brain Res.* **38**:233–249, 1972.

78. Björklund, A., Katzman, R., Stenevi, U., and West, K. A., Development and growth of axonal sprouts from noradrenaline and 5-hydroxytryptamine neurones in the rat spinal cord, *Brain Res.* **31**:21–33, 1971.

79. Moore, R. Y., Björklund, A., and Stenevi, U., Growth and plasticity of adrenergic neurons, in: *The Neurosciences*, Vol. 3 (F. O. Schmitt and F. G. Worden, eds.), pp. 961–977, MIT Press, Cambridge, Mass., 1974.

80. Murray, J. G., and Thompson, J. W., The occurrence and function of collateral sprouting in the sympathetic nervous system of the cat, *J. Physiol. (London)* **135**:133–162, 1957.

81. Guth, L., and Bernstein, J. J., Selectivity in the re-establishment of synapses in the superior cervical sympathetic ganglion of the cat, *Exp. Neurol.* **4**:59–69, 1961.

82. Liu, C. N., and Chambers, W. W., Intraspinal sprouting of dorsal root axons. Development of new collaterals and preterminals following partial denervation of the spinal cord in the cat, *Arch. Neurol. Psychiatry* **79**:46–61, 1958.

83. McCouch, G. P., Austin, G. M., Liu, C. M., and Liu, C. Y., Sprouting as a cause of spasticity, *J. Neurophysiol.* **21**:205–216, 1958.
84. Rose, J. E., Malis, L. I., Kruger, L., and Baker, C. P., Effects of heavy, ionizing, monoenergetic particles on the cerebral cortex. III. Histological appearance of laminar lesions and growth of nerve fibers after laminar destruction, *J. Comp. Neurol.* **115**:243–295, 1960.
85. Eccles, J. C., Eccles, R. M., and Shealy, C. N., An investigation into the effect of degenerating primary afferent fibers on the monosynaptic innervation of motoneurons, *J. Neurophysiol.* **25**:544–558, 1962.
86. Goodman, D.C., and Horel, J. A., Sprouting of optic tract projections in the brain stem of the rat, *J. Comp. Neurol.* **127**:71–88, 1966.
87. Raisman, G., Neuronal plasticity in the septal nuclei of the adult rat, *Brain Res.* **14**:25–48, 1969.
88. Lynch, G. S., Deadwyler, S., and Cotman, C. W., Postlesion axonal growth produce permanent functional connections, *Science* **180**:1364–1366, 1973.
89. Stenevi, U., Björklund, A., and Moore, R. Y., Morphological plasticity of central adrenergic neurons, *Brain Behav. Evol.* **8**:110–134, 1973.
90. Levi-Montalcini, R., and Chen, J. S., Selective outgrowth of nerve fibers *in vitro* from embryonic ganglia of *Periplaneta americana*, *Arch. Ital. Biol.* **109**:307–337, 1971.
91. Garber, B., and Moscona, A. A., Reconstruction of brain tissue from cell suspensions. I. Aggregation patterns of cells dissociated from different regions of the developing brain, *Dev. Biol.* **27**:235–243, 1972.
92. Barbera, A. J., Marchase, R. B., and Roth, S., Adhesive recognition and retinotectal specificity, *Proc. Natl. Acad. Sci. U.S.A.* **70**:2482–2486, 1973.
93. Chamley, J. H., Goller, I., and Burnstock, G., Selective growth of sympathetic nerve fibers to explants of normally densely innervated autonomic effector organs in tissue culture, *Dev. Biol.* **31**:362–379, 1973.
94. Olson, M. I., and Bunge, R. P., Anatomical observations on the specificity of synapse formation in tissue culture, *Brain Res.* **59**:19–33, 1973.
95. Blinzinger, K., and Kreutzberg, G., Displacement of synaptic terminals from regenerating motoneurons by microglial cells, *Z. Zellforsch. Mikrosk. Anat.* **85**:145–157, 1968.
96. Bernstein, J. J., and Bernstein, M. E., Neuronal alteration and reinnervation following axonal regeneration and sprouting in mammalian spinal cord, *Brain Behav. Evol.* **8**:135–161, 1973.
97. Gelfan, S., Field, T. H., and Pappas, G. D., The receptive surface and axonal terminals in severely denervated neruons within the lumbosacral cord of the dog, *Exp. Neurol.* **43**:162–191, 1974.

# INDEX